United States
Department
of Agriculture

Forest Service

**Rocky Mountain
Research Station**

General Technical
Report RMRS-GTR-275WWW

April 2012

Nonmarket Economic Values of Forest Insect Pests: An Updated Literature Review

Randall S. Rosenberger
Lauren A. Bell
Patricia A. Champ
Eric L. Smith

Rosenberger, Randall S.; Bell, Lauren A.; Champ, Patricia A.; Smith, Eric. L. 2012. **Nonmarket economic values of forest insect pests: An updated literature review.** Gen. Tech. Rep. RMRS-GTR-275WWW. Fort Collins, CO: U.S. Department of Agriculture, Forest Service, Rocky Mountain Research Station. 46 p.

ABSTRACT

This report updates the literature review and synthesis of economic valuation studies on the impacts of forest insect pests by Rosenberger and Smith (1997). A conceptual framework is presented to establish context for the studies. This report also discusses the concept of ecosystem services; identifies key elements of each study; examines areas of future research; and includes appendices that further explain nonmarket valuation methods, a narrative of each study, and tables that summarize each study. The primary services affected by insects are restricted in the literature to include recreation, aesthetic or scenic beauty of landscapes, and property values. Monetary metrics across studies include willingness to pay estimates per acre, per person or household, per tree, and/or for various levels of damages. While this literature is limited and heterogeneous, individual studies may prove useful in assessing current and future policies associated with forest insect pests in the United States.

Keywords: forest insect pests, nonmarket valuation, economic methods, ecosystem services, climate change

THE AUTHORS

Randall S. Rosenberger is an environmental economist at Oregon State University in the Department of Forest Ecosystems and Society, Corvallis, Oregon.

Lauren A. Bell is a graduate student at Oregon State University in the Applied Economics program, Corvallis, Oregon.

Patricia A. Champ is an economist for the USFS Rocky Mountain Research Station, Fort Collins, Colorado.

Eric L. Smith is with the Forest Health Technology Enterprise Group, Fort Collins, Colorado.

Cover photo: *Beetle-killed trees in Glacier National Park (William M. Ciesla, Forest Health Management International, Bugwood.org photograph).*

Available only online at http://www.fs.fed.us/rm/pubs/rmrs_gtr275.html.

CONTENTS

Introduction

Forests possess many components and processes that provide an array of ecosystem goods and services: timber, energy and water savings, pollution reduction, livestock forage, habitat for plants and animals, recreation opportunities, aesthetic landscapes, and biodiversity that enhance people's quality of life (Kline 2007). Forest resources also support local and regional economies through jobs and income generated from forestry, agriculture, tourism, and locational decisions of businesses, retirees, and others (Loomis 2002; Rosenberger and English 2005). The capability of a forest to provide these and many other ecosystem services and to maintain the quality of those ecosystem services depends on its health. A healthy forest is an essential component of a healthy ecosystem—a natural system that is capable of self-renewal, resilient in its response to disturbances (such as pest, fire, and other non-human and human-caused disturbances), and able to sustain the integrity of the natural and cultural benefits derived from it.

Many factors affect the health of a forest: air quality, fire, forest management practices and other human activities, wind, drought, disease, and insects. If managers want to take relevant tradeoffs into consideration, they need to be aware of the potential impacts that these disturbances may have on the quantity and quality of ecosystem services derived from forests. In addition, managers and other stakeholders need to understand the cause and effect of linkages between disturbances so that preventative action against one disturbance does not lead to the occurrence of another. Also, since ecosystems and forests are constantly changing, information on the impacts of these disturbances and their relationship to the health and sustainability of forest structures and processes is needed (Averill and others 1995).

Forest managers can weigh the tradeoffs associated with various management options in a number of ways. One method for assessing tradeoffs—economic benefit-cost analysis—is to compare the benefits of a management action with its costs. A complete benefit-cost analysis requires that all measurable benefits and costs be included. In the context of evaluating management actions to deal with forest threats, many of the benefits of a management action will be nonmarket goods and services. In a market setting, price is an indicator of the worth of a good representing its marginal value. Markets exist for some forest products, such as timber and livestock forage. However, other forest benefits, such as pollution reduction, aesthetic views, and some recreation opportunities, may not have markets, thus lacking prices as signals of their economic worth. These goods and services without markets are referred to as nonmarket goods and services. Nonmarket valuation techniques are used to estimate economic values for such goods.

The purpose of this report is to synthesize and evaluate the nonmarket valuation literature that quantifies the impacts of forest insect pests on ecosystem services (i.e., recreation, aesthetics, and homeowner benefits of property). The empirical estimates, economic measures of forest insect pest damages to nonmarket goods and services, reviewed in this report may be used in a benefit-cost analysis of forest pest management decisions/policies. Over the last four decades more than 20 nonmarket valuation studies pertaining to the economic effects of forest pests have been published (see Table 1). Forest diseases and pathogens have similar effects on forest health and derived economic values; however, limited valuation research was found for them. Therefore, we limit the scope of this review to forest insect pests.

Forest insect pests may directly affect forest commodities, such as timber, by damaging or killing trees, or they may indirectly affect non-commodity benefits, such as recreation experiences, by reducing the aesthetic appeal of areas where people recreate. Insects attack trees, causing discoloration of foliage, defoliation, or both, resulting in dead and down trees and visible damage to forests, which, in turn, may reduce the benefits derived from the forest and its products. They can also have negative impacts on the flow of ecosystem services provided by a forest. As with any damage, there is an associated cost or loss. The amount of this loss depends on a variety of factors, including the condition, type, and location of the forest, the magnitude of the outbreak, the kind of insect, the quality and intensity of the desired experience (aesthetic, recreational), and the scope of affected stakeholders. One important question facing forest managers is how many resources they should allocate toward the protection of forest health versus other management needs. The answer to this question depends partly on the level of physical damages that results from forest insect attacks. Another part of the answer depends on the economic value of these damages. The studies reviewed here, and their estimates of nonmarket values, can be used in an economic evaluative framework that enables forest managers to make more informed decisions.

Table 1—Economic studies of forest insect pests.

Study code	Author(s)	Year	Title
1	Payne and others	1973	Economic analysis of the gypsy moth problem in the Northeast: II. Applied to residential property
2	Wickman and Renton	1975	Evaluating damage caused to a campground by Douglas-fir tussock moth
3	Michalson	1975	Economic impact of mountain pine beetle on outdoor recreation
4	Moeller and others	1977	Economic analysis of the gypsy moth problem in the Northeast: III. Impacts on homeowners and managers of recreation areas
5	Leuschner and Young	1978	Estimating the southern pine beetle's impact on reservoir campsites
6	Walsh and Olienyk	1981	Recreation demand effects of mountain pine beetle damage to the quality of forest recreation resources in the Colorado Front Range
7	Walsh and others	1981a	Value of trees on residential property with mountain pine beetle and spruce budworm in the Colorado Front Range
8	Walsh and others	1981b	Appraised market value of trees on residential property with mountain pine beetle and spruce budworm damage in the Colorado Front Range
9	Loomis and Walsh	1988	Net economic benefits of recreation as a function of tree stand density
10	Walsh and others	1989	Recreational demand for trees in National Forests
11	Walsh and others	1990	Estimating the public benefits of protecting forest quality
12	Jakus and Smith	1991	Measuring use and nonuse values for landscape amenities: a contingent behavior analysis for gypsy moth control
13	Haefele and others	1992	Estimating the total value of forest quality in high elevation spruce-fir forests
14	Miller and Lindsay	1993	Willingness to pay for a state gypsy moth control program in New Hampshire: a contingent valuation case study
15	Holmes and Kramer	1996	Contingent valuation of ecosystem health
16	Thompson and others	1999	Valuation of tree aesthetics on small urban-interface properties
17	Haefele and Loomis	2001	Using the conjoint analysis technique for the estimation of passive use values of forest health
18	Kramer and others	2003	Contingent valuation of forest ecosystem protection
19	Asaro and others	2006	Control of low-level Nantucket pine tip moth populations: a cost benefit analysis
20	Holmes and others	2006	Exotic forest insects and residential property values
21	Huggett and others	2008	Forest disturbance impacts on residential property values
22	Price and others	2010	Insect infestation and residential property values: a hedonic analysis of the mountain pine beetle epidemic

Role of Forest Insects

Another concern that received little attention in the past economic studies is the role of forest insects in the natural processes of a forest. Forest insects are an integral part of the forest ecosystem (Anhold and others 1996). A healthy forest can sustain tree damages due to insects, and may even benefit from these disturbances. Forest pests act as natural thinning agents and can change the composition of the forest to provide more diversity and energy flow (e.g., increased sunlight penetration may increase understory growth, providing more forage for wildlife). The view that forest insects primarily cause negative impacts is based on a static view of a forest. When the dynamics of forest structures and processes are considered, forest insects may have an important and integral role in the complexity and overall health of a native forest. Therefore, economic models should consider the sustainability of the flow of ecosystem services from a forest in the form of products and services provided.

This report updates Rosenberger and Smith (1997) by reproducing and updating the literature review through summarizing the published economic valuation studies on the impacts of forest insect pests on nonmarket forest ecosystem services. It presents a conceptual framework to establish context for the economic valuation studies; discusses the concept of ecosystem services; identifies key elements of each study; provides a synthesis of the literature and areas of future research; and includes appendices that provide a narrative of each study and tables that summarize the studies. The scope of what constitutes a forest in the following studies ranges from large, publicly owned tracts of forested land to small, privately owned groups of trees (residential backyards). The services affected by insects are restricted in the literature to include recreation, aesthetic or scenic beauty of landscapes, timber production and property values. The economic values reported in the summaries of the studies are all given in the dollar value from the year of publication. Some of the estimates in a particular study are based on the worth of the property at the time of the study; therefore, it may not be correct to assume that property values have increased at the same rate as the price index.

The articles included in this study were screened by pre-determined criteria as a list of keywords that define the context of the review: forest pests (insects, diseases, and pathogens), nonmarket valuation, economics, forest fire, climate change, and forest management. An extensive literature search using various combinations of the keywords was conducted on databases such as Google Scholar, AgEcon Search, EconLit, RePEc, and EVRI. Only those studies conducted on forests in the United States were retained. This resulted in 34 articles that meet some or all of the keywords listed above. Ruling out all conceptual applications of economics and keeping only articles that contained an empirical model and numerical data narrowed the collection of articles. The final set of 22 articles are nonmarket economic valuation studies in the United States that include willingness to pay estimates obtained through the use of the travel cost, hedonic pricing, cost benefit analysis, contingent choice, or contingent valuation methods.

A Conceptual Framework

It is evident in the literature that there are many natural and human influences affecting the severity of forest pest outbreaks: fire, climate, and forest management. Knowledge of these influences and their relationship to forest pest populations can be useful when taking preventative action or controlling pest outbreaks and their impacts. The conceptual framework presented here provides a summary of the connections between forest pests and climate change, forest management, forest fire, and ecosystem services, using mountain pine beetle (*Dendroctonus ponderosae*) as an example. The intent of the framework is to provide context for the research presented in the economic valuation studies on the impacts of forest insect pests on nonmarket forest ecosystem services. Figure 1 displays our conceptual framework of how ecosystem services are affected by management and natural disturbances through their impacts on forest condition. Climate change overlays this framework in that it can directly and indirectly affect forest condition, management actions, and natural disturbances. Economic values derived from ecosystem services act as signals of social preferences and feed back into management action decisions.

Humans depend upon the Earth's ecosystems to provide goods and services that are critical for survival and the achievement of quality of life. The concept of ecosystem services was initially proposed as a way to describe the contributions of intact ecosystems to human well-being and advocate ecosystem protection (Kline 2007). Fisher and others (2008, 2009) and de Groot and others (2002) offer similar definitions of

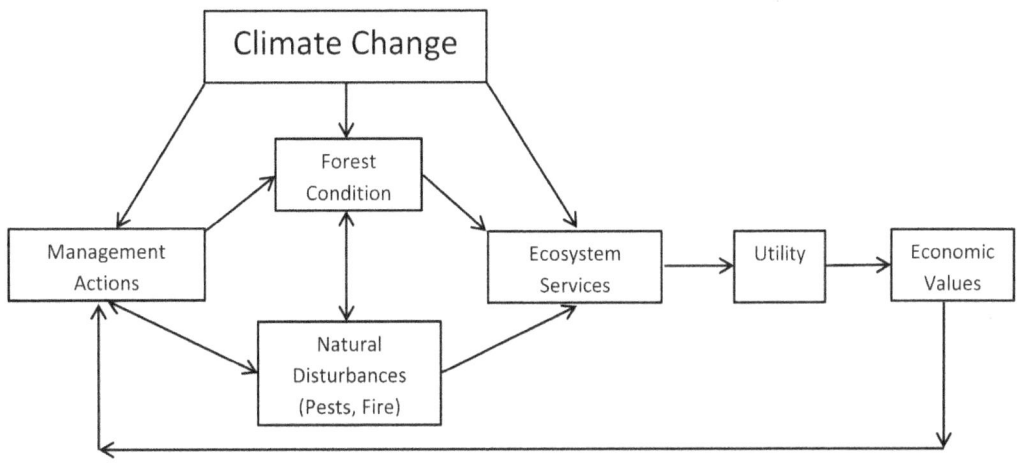

Figure 1—Conceptual framework of disturbances on ecosystem services and economic values.

ecosystem services: the natural processes and components of ecosystems utilized, directly or indirectly, to provide goods and services that satisfy human needs and produce human well-being. Defined in this way, ecosystem services are a product of ecosystem structures and processes. The functions, produced by the structures and processes of ecosystems, are conceptualized as ecosystem goods and services when human values are implied.

The composition, structure, and function of ecosystems have been rapidly altered throughout the history of human life on earth (Kremen and Ostfeld 2005). Natural disturbances such as pests and fire also directly and indirectly affect the delivery of ecosystem goods and services. For example, forest fires compete with timber and wood fiber industries, but they also affect recreational opportunities and the intrinsic value of a landscape, although these effects are naturally recoverable over time (Englin and others 2001). Similarly, pest outbreaks can directly and negatively affect the stock and flow of ecosystem services from forests; they also have indirect effects through their contributions to fire risk and release of stored carbon. Forest mortality caused by wildfire, insects, and diseases are principal sources of carbon emissions from forests in the western United States (Stephens 2005) and can lead to widespread loss of centuries' worth of carbon storage (Millar and others 2007).

The interactions between pests and their hosts affect the structures and processes of forest ecosystems. Thus, pests directly affect the forest's ability to perform ecosystem functions. When a change in ecosystem functioning occurs, it is valued in terms of the goods and services provided to humans. The value of a change in ecosystem services, caused by a forest pest outbreak, is dependent on the geographical context in which it occurs, including economic, social, and political factors (Kremen and others 2007). If the change in an ecosystem service is significant or the associated value is high, public policy and forest management may be impacted.

Pest populations are causally related to climate change, forest management practices, and forest fire. Climate change, forest management, and forest fire affect the composition of plants, the size of pest populations, and biotic and abiotic factors affecting both groups (Kremen and others 2007). Changes in landscape structure, caused by fire or human action, affect pest populations, host tree populations and their interactions at individual, community, and population scales (Kremen and others 2007). The population of many pests is affected by temperature, and the availability of susceptible host trees for food, reproduction, and overwintering. Similarly, the availability of host trees is affected by biotic and abiotic factors, forest management practices and forest fire.

In long-lived ecosystems such as forests, pests are often primary disturbance agents (Logan and others 2003). Forest pest outbreaks can impact the dynamics of ecosystems through tree morbidity and mortality, potentially over extensive areas (Kurz and others 2008). The mountain pine beetle, one of the most aggressive forest pests, is responsible for a considerable amount of tree mortality in North America (Carroll and others 2003; Romme and others 1986). Trees act as primary producers, carbon storage compartments, and support structures for other life and natural processes within forests. Tree mortality, caused by forest pests, can contribute to short term losses in timber production (Romme and others 1986), act as hazard trees in recreation areas (Walsh and Olienyk 1981), reduce the aesthetic beauty of landscapes (Jakus and Smith 1991), and depreciate the value of housing near outbreak areas (Price and others 2010). All of the estimates of willingness to pay to prevent negative effects on the aforementioned forest ecosystem services are positive (see Appendix C).

Dead trees may also serve as fuel for forests' second greatest disturbance agent—fire (Logan and others 2003). When a forest fire occurs, many of the forest ecosystem functions are damaged or destroyed. Functions such as nutrient and water regulation, carbon sequestration, food/raw material production, species composition, aesthetic information, recreation, and biodiversity can be disrupted (de Groot and others 2002; McCullough and others 1998). Ecosystem services produced by these functions include, but are not limited to, maintenance of local air quality; water flow timing, quantity and quality; influence on climate; building and manufacturing inputs; hunting and gathering of fish, game, and plant food sources; enjoyment of scenery; and travel in natural settings (de Groot and others 2002).

Mountain Pine Beetle

Forest pest disturbances are not always detrimental to the maintenance of ecosystem integrity (Logan and others 2003). The mountain pine beetle and other insects are an integral part of the forest ecosystem. Endemic populations of mountain pine beetle aid in the process of forest regeneration and the continuation of forest health. Romme and others (1986), citing Mattson and Addy (1975) and Moore and Hatch (1981), showed that outbreaks of mountain pine beetle, among other insects, led to only a brief drop in a forest's primary productivity. Over the short term (5-20 years), mountain pine beetle populations, at natural or outbreak levels, introduced more variation into long term primary production resulting in a more equitable distribution of biomass and resources among canopy, sub-canopy, and understory trees (Romme and others 1986). Beetles act as thinning agents, attacking larger, older trees that create openings for new growth and contribute fuel for eventual stand replacing fires (Anhold and others 1996). Flannigan and others (2000) showed the timing of the most active fire months was offset from the concentration of mountain pine beetle attacks by approximately a month (i.e., beetle attacks generally precede fires), potentially increasing fire size and intensity.

The concentrations of mountain pine beetle attacks generally occur in the late spring and in mature pine stands. Thus, the abundance of mature pine trees may regulate whether increasing temperatures, due to climate change, will promote mountain pine beetle outbreaks (Berg and others 2006). The survival of individual host trees depend, in part, on beetle-tree interactions and tree characteristics, such as size, age, bark thickness, phloem thickness, and resin pressure (Bradley and Tueller 2001). These characteristics that describe an individual tree's structure relate to its predisposition or susceptibility to mountain pine beetle attack. Large, slow growing trees are the preferred host of beetles due to their inability to resist the establishment of adults in the inner bark or phloem layer (Berg and others 2006). Keane and others (2002) argued that, at the stand level, the activities of forest insect pests and diseases are directly related to the stress and reduced vigor of tree species. As stand biomass and plant density increase, the competition among trees becomes more intense, causing increased stress. Management actions, such as fire suppression, and climate changes (e.g., in temperature and precipitation) may also stress trees and increase biomass, thus contributing to the intensity and location of beetle outbreaks.

For the past several decades, forest managers have used historical ecosystem conditions as the standard for assessing ecological sustainability, variability, and integrity. These ideals were used to set management goals for maintaining ecological systems and the goods, services, and amenities they provide (Lackey 1998; Millar and others 2007). However, in large part due to changes in climate, unprecedented rapid environmental changes are expected in the future (Millar and others 2007). Climate change, in particular global warming, may affect forest pest populations. The population dynamics of many forest pest species is dependent on temperature, which has a large effect on population size and distribution (Logan and others 2003). Changes in temperature and precipitation influence the survival, reproduction, and range of mountain pine beetle (Ayers and Lombardero 2000). In the western United States, warming annual temperatures have caused an upward shift in the range of acceptable habitat for mountain pine beetle. This has allowed the mountain pine beetle to populate high-elevation pine forests that were formerly uninhabitable (Kurz and others 2008). Global warming is projected to continue into the future and thus the range of the mountain pine beetle is expected to continue to expand (Kurz and others 2008). Likewise, the frequency, severity, and spatial distribution of spruce budworm outbreaks are expected to increase (Logan and others 2003).

Warmer summer temperatures brought on by global warming may decrease precipitation and lengthen the growing season (Millar and others 2006). Berg and others (2006), when comparing mountain pine beetle outbreaks to summer temperature, found that mountain pine beetle outbreaks occur after 5- to 6-year periods of warm summer temperatures. Their models estimated that, in Alaska, "the odds of a large outbreak occurring increased by 17.8 times (95% CI = 12.6-25.2) with each one-degree Celsius increase in average temperature" (p. 224), reaching 50% chance of occurrence at an average temperature of 10.3 degrees Celsius. The warm summer temperatures contribute to higher rates of mountain pine beetle reproduction and drought-induced stress on trees (Berg and others 2006; Millar and others 2006). First, the warmer temperatures during the late spring allow the mountain pine beetle to hatch, attack, and breed earlier in their life cycle, which induces a longer growing season for the larvae (Berg and others 2006). Second, direct solar radiation increases phloem temperatures in host trees and can as much as double the maturation rate of larvae, causing them to grow into adults in 1 year rather than 2 (Werner and Holsten 1985).

Warmer winter temperatures also affect the size and abundance of mountain pine beetle populations. If winters fail to reach the low temperatures historically consistent in the area, then winter survival of the mountain pine beetle increases in northern latitudes and in higher elevations. Direct measurements and records of wildland tree populations, cited by Ayers and Lombardero (2000), indicate that 100% mortality of the pine beetle occurs when air temperatures reach -16 degrees Celsius or less. This also implies that mountain pine beetles are living as far north as the winter temperatures will allow (Ayers and Lombardero 2000). Outbreaks of mountain pine beetle and other forest pests have also been attributed to local forest disturbances such as wind throw, floods, disease, drought, ice and snow damage, defoliation, landslides, and root disease (Logan and others 2003; Werner and others 1985). The occurrence of disturbances weakens tree stands creating hospitable breeding grounds for the insects.

Interactions between disturbances could be synergistic and may change rapidly with climate (Flannigan and others 2000). The potential future impacts of climate change on ecosystems have been estimated using climate and vegetation models, simulating two possible greenhouse gas emissions scenarios: high and low. These results show that temperatures across Oregon will increase 7 to 8.5 degrees Celsius by the end of the 21st century (Millar and others 2006). Patterns in precipitation, wind, and cloudiness are expected to change with a changing climate (Flannigan and others 2000). These scenarios reveal a 10 to 24% decrease in precipitation in the winter and a decrease of 10 to 40% in the summer in the Northwest (Millar and others 2006). While summers are already generally dry, the impact on winter precipitation may be significant and have exacerbating effects on pests, fires, and overall forest health. Thus, while forest management may help mitigate and adapt to factors affecting ecosystem services, climate change may exacerbate some of these efforts directly through changes in temperature and precipitation, and indirectly through increasing the frequency and severity of pest and fire disturbances. In order to evaluate the effects of climate change, forest fire, pests, and forest management practices on forest ecosystem services, information about the changes in the ecosystem services and their social value is required (Kline 2007). Describing changes in ecosystem services is often difficult due to lack of information about causal relationships, forest conditions, and policy and management activities. If measures of output changes exist, then the scope of inference is often limited and not transferable beyond a specific landscape (Kline 2007). Furthermore, without a rigorous and consistent classification system, double counting of ecosystem services benefits and confusion about what is being evaluated abound (Boyd and Banzhaf 2006; Brown and others 2007; Fisher and others 2008, 2009; Kline 2007).

The nonmarket valuation literature on forest insect pests is heterogeneous, as detailed below. While proxy measures are limited to a narrow set of ecosystem services (trees, views, recreation), monetary and other metrics are based on a wide range of units, including per acre, per tree, per person and per day, and vary based on level of impacts, including 3-7%, 15% and 30% damage in the near and/or far view. This heterogeneity in the literature prevents derivation of summary measures of monetary worth across studies for general use in assessing forest pest policies and practices. Instead, we suggest the reader use the summary tables below and detailed descriptions provided in the appendices to this report to identify a study or subset of studies relevant to his/her analysis context. If the reader finds no studies relevant to his/her needs, then we suggest he/she consider conducting an original study that directly meets his/her needs and adds to this body of research.

Key Elements of Economic Valuation Studies

There are many uses of the information contained in the economic valuation studies reviewed. The key elements identified in these studies help to define the context for the transfer or application of study outcomes to different contexts (Rosenberger and Smith 1997). These key elements also define the context of the data (e.g., region, forest type, kind of insect, magnitude of damages or infestation, visibility, and stakeholder groups represented), the valuation methodology used, choice of indicator variables, and the values at stake. The studies reviewed are listed in Table 1. The codes shown in the first column of Table 1 will be used to cross-reference the studies in Tables 2-8, and written and tabular summaries in Appendices B and C, respectively.

Forest Insect Pests Studied

The forest insect pests studied are listed in Table 2. Eight types of insects have been studied across the economic valuation literature, but they are not exhaustive of insects that impact forest health. Forest diseases and pathogens are little studied in the economic valuation literature. There are many types of diseases (rust and stem decay, fungi, mistletoe) and pathogens (seedling, root, wilt, canker, and foliage pathogens); still little is known about the magnitude or distribution of their effects (Kovacs and others 2011). Two notable exceptions include a valuation study on sudden oak death, published by Kovacs and others (2011) and Meldrum and others (2011) study of white pine blister rust. However, these recent studies on diseases and pathogens, along

Table 2—Forest insect pests studied.

Insect pest	Study code
Gypsy moth (*Lymantria dispar*)	1, 4, 12, 14, 17
Douglas-fir tussock moth (*Orgyia pseudotsugata*)	2
Mountain pine beetle (*Dendroctonus ponderosae*)	3, 6, 7, 8, 9, 10, 11, 22
Southern pine beetle (*Dendroctonus frontalis*)	5, 17
Spruce budworm (*Choristoneura occidentalis*)	7, 8, 11, 17
Balsam woolly adelgid (*Adelges piceae*)	13, 15, 18
Nantucket pine tip moth (*Rhyacionia frustrana*)	19
Hemlock woolly adelgid (*Adelges tsuga*)	20, 21
Unidentified	16

with a myriad of other disturbances both natural (e.g., forest fire, wind throw, climate change, and drought) and human induced (e.g., silvicultural practices, pollution, and forest fire prevention), are not included in this review as we focus on forest insect pests.

The gypsy moth, *Lymantria dispar*, is an introduced pest that originated in Europe and Asia. The most common host species for the gypsy moth are oak and aspen trees (Liebhold 2003). The gypsy moth has little effect on commercial tree stands, but imposes a large impact on ornamental trees and trees in recreation areas (Haefele and Loomis 2001).

The Douglas fir tussock moth, *Orgyia pseudotsugata*, is a native defoliator of fir trees in western North America. Preferred hosts include Douglas fir, white fir, and grand fir trees. Severe outbreaks, which develop explosively, have occurred from British Columbia to Arizona and New Mexico. Defoliation caused by the Douglas fir tussock moth can kill or weaken many trees. Weakened trees suffer from retarded growth and can eventually be killed by bark beetles (Wickman and others 1998).

The mountain pine beetle, *Dendroctonus ponderosae*, is an insect native to the forests of western North America (Kurz and others 2008). Populations of mountain pine beetle can be found from Alaska and the Yukon Territory to the southwestern United States (Berg and others 2006). It is the most aggressive of the bark beetles found in western North America where it primarily infests and kills lodge pole pine, ponderosa pine, white bark pine, and limber pine (Jenkins and others 2008).

The southern pine beetle, *Dendroctonus frontalis*, is a native insect in the southeastern United States. It can attack any species of pine, but is found most commonly in loblolly, shortleaf, Virginia, pond, and pitch pines. Its range extends from Pennsylvania to Texas where it impacts commercial pine stands and wilderness areas (Haefele and Loomis 2001).

The spruce budworm, *Choristoneura occidentalis*, is native to most fir stands in the western United States. The spruce budworm is the most widely distributed defoliating insect in this region, impacting commercial Douglas and grand fir stands in the Pacific Northwest.

The balsam woolly adelgid, *Adelges piceae*, is an introduced species from Europe. It is considered to be a serious pest to the Christmas tree industry, seed production, and forested landscapes. The balsam woolly adelgid is distributed throughout the United States, but found primarily in the Pacific Northwest and the southern Appalachian Mountains (Michigan DNRE 2001).

The Nantucket pine tip moth, *Rhyacionia frustrana*, is a native forest insect pest that ranges from Massachusetts to Florida and as far west as Texas. The Nantucket pine tip moth is most damaging to pine plantations and to wild pine seedlings in open areas. It poses problems because of forestry trends that favor the establishment of large areas of pine plantations (Asaro and others 2006).

The hemlock woolly adelgid, *Adelges tsuga*, was introduced to the United States from Japan in the 1950s. It causes mortality among eastern and Carolina hemlocks. The hemlock woolly adelgid affects both ornamental hemlocks and hemlock forests in the Northeast, mid-Atlantic, and the South (Holmes and others 2006).

Regions Studied

The regions where the economic studies were conducted are identified in Table 3. The majority of the studies were conducted in the mountainous West and Northeast. The mountainous West region was of

Table 3—Regions of the United States where economic studies of forest insect pests were conducted.

Region	Study code
Mountainous West	6, 7, 8, 9, 10, 11, 22
Northeast	1, 4, 12, 14, 17, 20, 21
Northwest	3, 17
South	5, 19
Southeast	13, 15, 17, 18
Southwest	2, 16

particular interest to the USDA Forest Service in the late 1970s and early 1980s when a project series was published assessing the impacts of mountain pine beetle and western spruce budworm in the Colorado Front Range (Walsh and Olienyk 1981; Walsh and others 1981a, b). Five studies were conducted in the Northeast region to measure the economic impacts of the gypsy moth and, most recently, two studies assessing the impacts of the hemlock woolly adelgid on property values (Holmes and others 2006; Huggett and others 2008). The remaining studies investigate the Douglas-fir tussock moth in the Southwest, the mountain pine beetle and spruce budworm in the Northwest, the southern pine beetle and Nantucket pine tip moth in the South, and the balsam woolly adelgid in the Southeast.

Land-Use Areas Studied

The type of land-use designation directly influences the magnitude and type of impact a forest insect pest can have on an area. The three land-use designations are urban, wildland, and the wildland-urban interface (i.e., "areas where human-made developments are in proximity to or intermingle with undeveloped wilderness" (Price and others 2010, p. 417)), as listed in Table 4.

The extent of the damage, measured as monetary losses, caused by the insects is expected to be greater in the urban and wildland-urban interface areas (Rosenberger and Smith 1997). This is because the amount of damage caused by the insects is more visible than in forested wildland areas, thus directly affecting a larger body of stakeholders. Property values in urban and wildland-urban interface areas can be greatly affected by tree damage and mortality. Moreover, as established in Holmes and others (2006), relationships show that tree health on both individual and neighboring properties matter. As the insect damage becomes more distant, the negative effect on property values diminishes. Insect damage will eventually fade through the establishment and growth of young trees, but the effect it has on recreation, aesthetic values, and property values in the short run can be considerable. Some of the studies are concerned with the value added by

Table 4—Forest insect pest studies by designated land use area.

Land-use area	Study code
Urban	1, 4, 12, 20, 21
Wildland	2, 3, 5, 6, 9, 10, 11, 13, 15, 17, 18, 19
Wildland-urban interface	7, 8, 14, 16, 22

trees to recreation activities, property values, and aesthetic viewsheds. Others are primarily concerned with forest-dominated wildland areas, where the supply of recreation and forest commodities are most affected (Rosenberger and Smith 1997).

Stakeholders

The way forest ecosystems are managed is of great importance, as is the recognition of the relevant stakeholders to the area. The values that are attributed to an ecosystem, and the goods and services it provides, depend upon the stakeholders benefiting from these services (Hein and others 2006). A stakeholder is any group or individual who can affect or is affected directly or indirectly by forest ecosystem services and the changes in those services as a result of an insect infestation (Hein and others 2006; Rosenberger and Smith 1997). When formulating any management plan, the stakeholders should always be considered, if not included, in the decision making process. Table 5 lists the stakeholders identified in the economic studies. Forest managers, planners, and decision makers are implicit throughout the list.

Table 5—Stakeholders identified in forest insect pest studies.

Stakeholder	Study code
General public	13, 14, 15, 17, 18
Homeowners	1, 4, 7, 12, 16, 20, 21, 22
Land managers	4, 19
Real estate appraisers	8
Recreationists	2, 3, 5, 6, 9, 10, 11, 13, 15

Values Estimated

Recreation is the most commonly identified value in the studies. Other values that are identified include aesthetics, passive-use (including option, existence, and bequest values of forest health), property, and total value (the sum of use and passive use values) (see Table 6). Over the past few decades the United States has experienced rapid growth in communities with recreational opportunities and natural amenities (Price and others 2010). The result has been a dramatic increase in development in the wildland-urban interface. The more recent studies focus on the contributed residential property value of trees and the total value of forests. The economic estimates included in the studies are primarily short-term responses to a static comparison of forest conditions. Therefore, these estimates do not necessarily include the social, cultural, and ecological importance of forest functioning, such as maintaining biodiversity, sustaining natural processes, and providing historic cultural identity (Rosenberger and Smith 1997).

Table 6—Values identified and measured in economic studies of forest insect pests.

Value	Study code
Aesthetics	2, 12
Passive-use	11, 13, 15, 17
Property	1, 7, 8, 16, 20, 21, 22
Recreation	2, 3, 4, 5, 6, 9, 10, 11, 13, 15
Total Value	11, 13, 14, 15, 18, 19

Table 7—Nonmarket valuation methods used in economic studies of forest insect pests.

Nonmarket valuation method	Study code
Contingent valuation	6, 7, 8, 9, 10, 11, 12, 13, 14, 15, 18
Contingent choice	17
Cost benefit	19
Hedonic property	1, 16, 20, 21, 22
Travel cost	3, 5, 6, 10
Other	2, 4

Table 8—Forest insect impact indicator variables used in economic studies.

Impact indicator	Study code
Number of trees	1, 2, 6, 7, 8, 9, 10, 11, 17, 22
Presence of insect	4, 18
Tree size	6, 7, 8, 9
Visible damage	2, 3, 4, 6, 7, 12, 13, 14, 15, 19, 20, 21
Other	5, 6, 7, 8, 16

Nonmarket Valuation Methods Used

Table 7 shows the different types of nonmarket valuation methods used in the studies. Contingent valuation is the most commonly used method. It is the most versatile of the nonmarket valuation methods and can measure recreation, property, and aesthetic benefits. The travel cost method was used in four studies. The hedonic pricing method is used in five studies, all of which measure property value. Other methods used include contingent choice, cost benefit, and replacement and expenditure cost estimation methods. For further information about each of the nonmarket valuation methods please see Appendix A.

Forest Insect Impact Indicators Used

The visible effects of forest insect pests vary between species and the hosts on which they prey. The manner in which insects can damage a forest include dead and down trees, the density of the forest, and amount of visible damage. Table 8 shows that the most common insect impact indicators used in these studies are visible damage and number of trees. When insects attack trees, they damage and sometimes kill them, resulting in fewer trees per acre and, in the short term, increased evidence of damage (Rosenberger and Smith 1997). All of the impact indicators used are static, only considering forest conditions at a given point in time and excluding long-term effects and outcomes.

Discussion/Synthesis of the Literature

Overall the reviewed studies show that people value many attributes, including quality, of forests. Trees add to the market value of homes, enhance recreational experiences, and provide a host of ecosystem services. Forest pests, especially when in outbreak status, can substantially affect forest quality and the flow of ecosystem services provided by them. Thus, forest pest outbreaks negatively affect property values, recreational experiences, and the provision of ecosystem services. Both public and private lands are affected by insect damages, especially in areas where there are large numbers of users and passers-by, such as campgrounds, urban parks, and scenic greenways. Decision makers are interested stakeholders in determining the relevant extent of insect damage mitigation to undertake. Land-use zones include urban, wildland-urban, and wildland areas.

The techniques employed in the economic estimation of these nonmarket damages (benefits) of pest infestations (control programs) are the contingent valuation, travel cost, and hedonic pricing methods. Also employed are direct cost estimates through the estimation of replacement costs and financial losses. The relevant metrics used in these measurements include dollars (market prices, expenditures, willingness to pay, or consumer surplus) and recreation days. The estimated models for the relevant studies identify several indicators of pest impacts. These include number of trees per acre, percentage of visible damage (e.g., dead and down trees, defoliation rates, and discoloration of foliage), physical presence of insects, size of trees, and percentage of tree species composition. All indicators were found to be positively related to the benefits generated, with the exception of tree species composition, which may be either positively or negatively related. In other words, the negative impacts of pest infestations on the level of the indicator variables predominantly result in a decrease in benefit derived from the resource (or conversely, result in an increase in the level of damages).

While the literature provides useful nonmarket valuation data for forest management decision making in the United States, it is limited (see areas of future research in the next section). Constraints to the ultimate usefulness of the values found within the literature will be the comprehensiveness and quality of the body of literature (Johnston and Rosenberger 2010; Pendleton and others 2007). Collectively, the studies reviewed cover a wide range of forest pests, affected values (property, direct and indirect use, and passive use—bequest and existence—values), and use a variety of nonmarket valuation methods. However, many of the studies are concentrated in specific areas of the United States that do not reflect the wide range of the pests. Many of the studies are also dated, leading to concerns about the transferability of their estimates for current management and policy assessment. The context of the studies is also limited in that they deal with outbreak conditions of the various pests—pests as a natural forest disturbance are largely unexplored. Thus, while some generalizations may be supported by this literature, the age of some studies, spatial extent and intensities of disturbances evaluated, and the geographic location of studies may limit their general use in benefit-cost analyses. Furthermore, most studies do not suggest managerial responses to the disturbances or mitigation strategies to reduce the real and perceived damages caused by them or evaluate confounding factors such as climate change, fire, and development patterns. At best an analyst may find a single study or subset of studies that match well with the policy context being evaluated.

The mountain pine beetle is the most studied forest pest in the literature. Four of the mountain pine beetle studies also looked at the spruce budworm on four occasions. Studies that include the mountain pine beetle and other pests evaluated their effects on recreation (Loomis and Walsh 1988; Michalson 1975; Walsh and

Olienyk 1981; Walsh and others 1989, 1990), property values (Price and others 2010; Walsh and others 1981a, b), passive use values, and total value (the sum of use and passive use values) (Walsh and others 1990). All but one of the mountain pine beetle studies (Michalson 1975) were conducted on the Eastern Front Range of Colorado's Rocky Mountains, despite the fact that their habitat ranges from Alaska and the Yukon Territory down to the southwestern United States (Berg and others 2006). The studies estimating the economic effects of mountain pine beetle on recreation all occur in wildland designated use areas and share the same affected stakeholders—recreators. Mountain pine beetle studies appear consistently in the literature with an increase during the late 1970s and early 1980s, due to increased interest of the USDA Forest Service.

The economic literature reveals that there is a negative relationship between the effects of forest pests and the quality and visitation rate for recreation. These negative effects are associated with visible damages (Haefele and others 1992; Holmes and Kramer 1996; Kramer and others 2003; Michalson 1975; Walsh and Olienyk 1981), tree density (Leuschner and Young 1978; Loomis and Walsh 1988; Walsh and Olienyk 1981; Walsh and others 1989, 1990; Wickman and Renton 1975), and tree size (Loomis and Walsh 1988; Walsh and Olienyk 1981). Pest impacts on recreation lead to a reduction in the quality of the recreation experience measured through decreased consumer surplus, but also through a decrease in the number of visits.

Pests kill or cause visible damage to trees, such as color change or defoliation. Residential properties are visually affected by mountain pine beetle and other pests in the near and far view, which has an effect on the satisfaction of owning and living on the property in the short run. This change in satisfaction is reflected in the property's value (Holmes and others 2006; Huggett and others 2008; Price and others 2010; Walsh and others 1981a). Forest pests are associated with decreases in property value (both developed and undeveloped). Healthy trees add positive value to a property through aesthetic value, ecosystem services, and increased owner satisfaction (Huggett and others 2008; Price and others 2010). Trees are so important to a property's value that homeowners are willing to pay to protect the trees from pests on their property and in the viewshed of their homes (Jakus and Smith 1991; Miller and Lindsay 1993; Payne and others 1973; Price and others 2010; Walsh and others 1981a, b).

The literature also shows that nonuse or passive-use benefits are more than three-and-a-half times greater than recreation-use benefits (Holmes and Kramer 1996;

Kramer and others 2003; Walsh and others 1990). Walsh and others (1990) emphasize the importance of measuring passive use values when calculating the total value of forest ecosystem services. Without the inclusion of passive use values, an economic assessment for management or policy would understate the true value of a forest and its realized ecosystem services. Without adequate information, a sub-optimal conclusion or decision could be reached.

Other forest pests identified in this review (Douglas fir tussock moth, southern pine beetle, Nantucket pine tip moth, and the Hemlock wooly adelgid) were not represented enough by the literature to draw conclusions or make comparisons beyond the individual articles, although the same general trends and conclusions of the broader literature are supported by them (see Appendix B for the summaries of the articles).

Conclusions and Areas of Future Research

This literature review shows evidence that insects cause significant economic damage beyond the damage to market commodities, at least in the short term. However, estimates of potential economic damages due to pest outbreaks reported in this literature are mostly site-specific and species-specific, providing a thumbnail sketch of the possible benefits from control programs or policies. The transferability of the economic measures and models are contingent on the context of these original studies and their correspondence with characteristics of the target site or policy in a benefit transfer (Johnston and Rosenberger 2010; Rosenberger and Loomis 2003; Rosenberger and Phipps 2007). Further research in the area of the economic impacts of insects and other disturbance agents on the productive capability and sustainability of different environmental resources or areas will add to the stock of knowledge and help guide future decisions and policies.

Confidence in our ability to transfer the stock of knowledge may decay over time. As noted by Pendleton and others (2007), "for values to be relevant to current policy-making, they need to reflect current estimates of nonmarket values" (p. 370). The economic valuation literature on the effects of forest pests has an equally distributed publication rate between 1973 and 2010. However, mountain pine beetle valuation studies peaked during the 1980s, with most estimates being at least 10 years old. Spruce budworm studies also peaked in the 1980s and have since only been studied by Haefele and Loomis (2001). Studies

valuing the effects of the gypsy moth dropped off in the 1980s and 2000s, with the last study occurring in 1993. The first of the balsam wooly adelgid valuation studies was reported in 1992 (Haefele and others 1992) and there have only been two studies reported since then. The literature is thin for the economic effects of insects such as the Douglas-fir tussock moth, southern pine beetle, Nantucket pine tip moth, and hemlock wooly adelgid. This implies that any attempt to use benefit transfer in valuing the current economic effects of forest pests would potentially rely on older information. Since methods for nonmarket valuation have been critiqued and updated, stakeholder preferences may have changed, and the scale and intensity of outbreaks may have increased over time, dependence on older data may affect, whether real or perceived, the accuracy and relevance of values obtained through benefit transfer (Johnston and Rosenberger 2010; Pendleton and others 2007). Therefore, not only do we need to expand our stock of knowledge through new primary research, we also need to replicate or verify older results.

All species of forest pests have long histories of interaction with their environments (Logan and others 2003). Still, little is known about how stakeholders perceive forest pest outbreaks based on their prior knowledge of historic and natural patterns of pest damages, and whether damage levels are based on human or natural disturbances. Information could include a general history of the insect affecting the area; the role the insect plays in the functioning of the ecosystem; and human actions that affect the extent and intensity of an outbreak. There is also a need to research whether this information changes people's attitudes toward pest outbreaks and the amounts people are willing to pay to avoid or mitigate them. Furthermore, stakeholders' attitudes toward different management strategies and their appreciation of derived ecosystem services may directly affect their willingness to pay for pest related programs and policies.

Most of the reviewed valuation studies are static, reflecting a point in time. An issue in need of further inquiry is the role static measures of economic damages may have in a larger, dynamic ecosystem management context. Static economic analysis does not estimate the economic worth of natural processes or ecosystem functions that define healthy forests and ecosystems. However, when management objectives converge (ecological, economic, and social), static measures may themselves be indicators of the human component of forest health management.

Most of the studies included in this review gather data and generate estimates for relatively small geographical areas. The range of many forest pests extends well past the study areas in the literature. For example, the mountain pine beetle can be found from Alaska all the way to the southwestern United States, yet the literature is concentrated on the Colorado Front Range of the Rocky Mountains (Walsh and Olienyk 1981; Walsh and others 1981a, b, 1989, 1990; Loomis and Walsh 1988; Price and others 2010). There are still many areas affected by mountain pine beetle and other pests that remain unstudied. Given this discrepancy, there is a need for nonmarket valuation studies to be done on broad spatial scales (i.e., over the entire range of a pest species). If studies were conducted on a broader scale, the estimates would be more appropriate for policy and management scale decision making.

Furthermore, scientific understanding of the effects of forest pests could benefit from further study of the spatial and temporal dimensions of pest outbreaks. Knowledge of the extent of an outbreak as well as the time scale over which the outbreak will affect forest ecosystem functions and services can improve economic estimates and policy decision making. In particular, the studies listed in this review provide site specific data and estimates, while policy making often occurs on a regional, state, or national scale, and is in effect for a number of years.

The studies in this review have valued forest pest outbreaks in urban (Holmes and others 2006; Huggett and others 2008), wildland (Asaro and others 2006; Haefele and Loomis 2001; Kramer and others 2003), and wildland-urban interface settings (Price and others 2010; Thompson and others 1999). Still, it is unknown whether peoples' willingness to pay and support management alternatives changes based on the type of land use at the location of the outbreak. Wildland forest pest outbreaks affect ecosystem functions and services differently than wildland-urban interface outbreaks or urban outbreaks. Thus, it is expected that people value the effects of the outbreaks differently. The issue then becomes identifying what people are willing to pay to avoid the same forest pest outbreak (identical in severity, spatial, and temporal scale) in the three land use designations and comparing the results.

The abundance of mature trees plays a large role in the ability of mountain pine beetle and other forest pests to reach outbreak levels (Berg and others 2006). Thus, any measures that forest and property managers take to maintain their properties are likely to affect mountain pine beetle and other pest populations. Catastrophic pest outbreaks in the Colorado Front Range (Price and others 2010) and elsewhere over the past three decades have forced some urgency among decision makers,

scientists, and the general public to deal with the changes in ecosystem services (Rosner 2009; Spyratos and others 2004). The literature to date weakly addresses the use of adaptive management (i.e. silviculture, stand rotation, and age composition) to control for or lessen the probability of pest outbreaks' scales and intensities. A comprehensive understanding of the relative influence of host species and climate on pest outbreaks over homogeneous regions is necessary to predict how frequently and with what severity pest outbreaks will occur. Lessons can be learned from the literature on fire and adopted prevention and mitigation strategies in each land use classification. For example, fire spread models using housing and vegetation data have been used to predict fire size and probability distributions (Spyratos and others 2004), and 'at risk' houses and communities have been targeted by informational campaigns to reduce or minimize risks to property through fire-resistant landscaping and construction materials. Thus, it seems reasonable that similar models and management strategies could be developed for forest pests.

Every pest outbreak has many contributing factors. Some of the factors can be controlled by human intervention while others are controlled purely by ecosystem functions. Therefore, there exist elements of uncertainty and risk in the management of forest ecosystems. Risk and uncertainty of events occurring, and their relative spatial and temporal impacts, need to be better understood in the case of forest pests. Scientific measures of risk and lay perceptions of risk often diverge, which is why it is important to inform decision makers and the public of the risk and uncertainty of potential forest pest outbreaks including likelihoods, confidences, and ranges of uncertainty. Better means of reporting and communicating actual risk levels can help inform the voting public as well as influence socially constructed perceptions of risk (Rosenberger 2006).

This review was restricted to forest insect pests due to the limited focus of nonmarket valuation studies on diseases and pathogens—notable exceptions include Kovacs and others (2011) and Meldrum and others (2011). It should not be construed that diseases and pathogens are less important or result in lower damages than forest insect pests. Forest diseases and pathogens as disturbance agents may result in similar physical damages to forest ecosystems. Thus, until primary valuation research focuses on diseases and pathogens, the valuation literature on forest insect pests, as reviewed here, may be used to inform forest policy and management of disease and pathogen prevention and mitigation—although the validity of such transfers of information remains an empirical question.

References Cited

Anhold, J.A.; Jenkins, M.J.; Long, J.N. 1996. Management of lodgepole pine stand density to reduce susceptibility to mountain pine beetle attack. Western Journal of Applied Forestry. 11(2): 50-53.

Arrow, K.; Solow, R.; Portney, P.; Leamer, E.; Radner, R.; Schuman, H. 1993. Report of the NOAA panel on contingent valuation. Federal Register. 58(10): 4602-4614.

Asaro, C.; Carter, D.R.; Berisford, C.W. 2006. Control of low-level Nantucket pine tip moth populations: A cost benefit analysis. Southern Journal of Applied Forestry. 30(4): 182-187.

Ayers, M.P.; Lombardero, M.J. 2000. Assessing the consequences of global change for forest disturbance from herbivores and pathogens. The Science of the Total Environment. 262(3): 263-286.

Berg, E.E.; Henry, J.D.; Fastie, C.L.; De Volder, A.D.; Matsuoka, S.M. 2006. Spruce beetle outbreaks on the Kenai Peninsula, Alaska, and Kluane National Park Reserve, Yukon Territory: Relationship to summer temperature and regional differences in disturbance regimes. Forest Ecology and Management. 227(3): 219-232.

Boyd, J.; Banzhaf, S. 2007. What are ecosystem services? The need for standardized environmental accounting units. Ecological Economics. 63(2-3): 616-626.

Bradley, T.; Tueller, P. 2001. Effects of fire on bark beetle presence on Jeffery pine in the Lake Tahoe Basin. Forest Ecology and Management. 142(1-3): 20-214.

Brookshire, D.S.; Thayer, M.A.; Schulze, W.D.; d'Arge, R.C. 1982. Valuing public goods: a comparison of survey and hedonic approaches. The American Economic Review. 72(1): 165-177.

Brown, T.C.; Bergstrom, J.C.; Loomis, J.B. 2007. Defining, valuing, and providing ecosystem goods and services. Natural Resources Journal. 47(1): 329-376.

Carroll, A.L.; Taylor, S.W. [and others]. 2003. Effects of climate change on range expansion by the mountain pine beetle in British Columbia. In: Shore, T.L.; Brooks, J.E.; Stone, J.E., comps. Proceedings: Mountain Pine Beetle Symposium: Challenges and Solutions; 2003 October 30-31; Kelowna, British Columbia. Information Report BC-X-399. Victoria, BC: Natural Resources Canada, Canadian Forest Service, Pacific Forestry Centre. 298 p.

Champ, P.; Boyle, K.; Brown, T. 2003. A primer on nonmarket valuation. Dordrecht, The Netherlands: Kluwer Academic Publishers. 592p.

de Groot, R.S.; Wilson, M.A.; Boumans, R.M.J. 2002. A typology for the classification, description and valuation of ecosystem functions, goods and services. Ecological Economics. 41(3): 393-408.

Englin, J.; Loomis, J.; Gonzalez-Caban, A. 2001. The dynamic path of recreational values following a forest fire: A comparative analysis of states in the Intermountain West. Canadian Journal of Forest Research. 31(10): 1837-1844.

Flannigan, M.D.; Stocks, B.J.; Wotton, B.M. 2000. Climate change and forest fires. Science of the Total Environment. 262(3): 221-229.

Fisher, B.; Turner, R.K.; Morling, P. 2009. Defining and classifying ecosystem services for decision making. Ecological Economics. 68(3): 643-653.

Fisher, B.; Turner, K.; Zylstra, M.; Brouwer, R.; [and others]. 2008. Ecosystem services and economic theory: Integration for policy-relevant research. Ecological Applications. 18(8): 2050-2067.

Freeman, A.M., III. 1979. Hedonic prices, property values and measuring environmental benefits: a survey of the issues. Scandinavian Journal of Economics. 81: 154-171.

Haefele, M.; Kramer, R.A.; Holmes, T. 1992. Estimating the total value of forest quality in high-elevation spruce-fir forests. In: Payne, C.; Bowker, J.M.; Reed, P.C., comps. The economic value of wilderness: proceedings of the conference; 1991 May 8-11; Jackson, WY. Gen. Tech. Rep. SE-78. Asheville, NC: U.S. Department of Agriculture, Forest Service, Southeastern Forest Experiment Station. pp. 91-96.

Haefele, M.A.; Loomis, J.B. 2001. Using the conjoint analysis technique for the estimation of passive use values of forest health. Journal of Forest Economics. 7(1): 9-24.

Hein, L.; van Koppen, K.; de Groot, R.S.; van Ierland, E.C. 2006. Spatial scales, stakeholders and the valuation of ecosystem services. Ecological Economics. 57(2): 209-228.

Huggett Jr., R.J.; Murphy, E.A.; Holmes, T.P. 2008. Forest disturbance impacts on residential property values. In: Holmes, T.P.; Prestemon, J.P.; Abt, K.L., eds. The economics of forest disturbances: Wildfires, storms, and invasive species. North Carolina: Springer Science. pp. 209-228.

Holmes, T.P.; Kramer, R.A. 1996. Contingent valuation of ecosystem health. Ecosystem Health. 2(1): 1-5.

Holmes, T.P.; Murphy, E.A.; Bell, K.P. 2006. Exotic forest insects and residential property values. Agricultural and Resource Economics Review. 35(1): 155-166.

Jakus, P.; Smith, V.K. 1991. Measuring use and nonuse values for landscape amenities: a contingent behavior analysis of gypsy moth control. Discussion paper QE92-07. Washington, DC: Resources for the Future. 48 p.

Jenkins, M.J.; Hebertson, E.; Page, W.; Jorgensen, C.A. 2008. Bark beetles, fuels, fires and implications for forest management in the Intermountain West. Forest Ecology and Management. 254(1): 16-34.

Johnston, R.J.; Rosenberger, R.S. 2010. Methods, trends and controversies in contemporary benefit transfer. Journal of Economic Surveys. 24(3): 479-510.

Keane, R.E.; Ryan, K.C.; Veblen, T.T.; Allen, C.D.; Logan, J.; Hawkes, B. 2002. Cascading effects of fire exclusion in Rocky Mountain ecosystems: A literature review. Gen. Tech. Rep. RMRS-GTR-91. Fort Collins, CO: U.S. Department of Agriculture, Forest Service, Rocky Mountain Research Station. 24 p.

King, D.M.; Mazzotta, M.J. 2000. *Ecosystem valuation* [Homepage of Contingent Choice Method], [Online]. Available: http://ecosystemvaluation.org/contingent_choice.htm [12 September 2010].

Kline, J.D. 2007. Defining an economics research program to describe and evaluate ecosystem services. Gen. Tech. Rep. PNW-GTR-700. Portland, OR: U.S. Department of Agriculture, Forest Service, Pacific Northwest Research Station. 46 p.

Kovacs, K.; Holmes, T.P.; Englin, J.E.; Alexander, J. 2011. The dynamic response of housing values to a forest invasive disease: Evidence from a sudden oak death infestation. Environmental and Resource Economics. 49(3): 445-471.

Kramer, R.A.; Holmes, T.P.; Haefele, M. 2003. Contingent valuation of forest ecosystem protection. In: Sills, E.O.; Abt, K.L., eds. Forests in a market economy. Boston: Kluwer Academic Publishers. pp. 303-320.

Kremen, C.; Ostfeld, R.S. 2005. A call to ecologists: measuring, analyzing, and managing ecosystem services. Frontiers in Ecology and the Environment. 3(10): 540-548.

Kremen, C.; Williams, N.M.; Aizen, M.A. [and others]. 2007. Pollination and other ecosystem services produced by mobile organisms: a conceptual framework for the effects of land use change. Ecology Letters. 10(4): 299-314.

Kurz, W.A.; Dymond, C.C.; Stinson, G.; Rampley, G.J.; Neilson, E.T.; Carroll, A.L.; Ebata, T.; Safranyik, L. 2008. Mountain pine beetle and forest carbon feedback to climate change. Nature. 452(7190): 987-990.

Lackey, R.T. 1998. Seven pillars of ecosystem management. Landscape and Urban Planning. 40(1): 21-30.

Leuschner, W.A.; Young, R.L. 1978. Estimating the southern pine beetle's impact on reservoir campsites. Forest Science. 24(4): 527-537.

Logan, J.A.; Régnière, J.; Powell, J.A. 2003. Assessing the impacts of global warming on forest pest dynamics. Frontiers in Ecology and the Environment. 1(3): 130-137.

Loomis, J.B. 2002. Integrated public lands management: Principles and applications to national forests, parks, wildlife refuges, and BLM lands. Second edition. New York, NY: Columbia University Press. 544 p.

Loomis, J.B.; Walsh, R.G. 1988. Net economic benefits of recreation as a function of tree stand density. In: Schmidt, W.C., ed. Proceedings: future forests of the mountain west: a stand culture symposium; 1986 September 29-October 3; Missoula, MT. Gen. Tech. Rep. INT-243. Ogden, UT: U.S. Department of Agriculture, Forest Service, Intermountain Research Station. pp. 370-375.

Loomis, J.; Walsh, R. 1997. Recreation economic decisions: Comparing benefits and costs. 2nd Edition. State College, PA: Venture Publishing. 440 p.

Mattson, W.J.; Addy, N.D. 1975. Phytophagous insects as regulators of forest primary production. Science. 190(4214): 515-522.

McCullough, D.G.; Werner, R.A.; Neumann, D. 1998. Fire and insects in northern boreal forest ecosystems of North America. Annual Review of Entomology. 43(1): 107-127.

Meldrum, J.R.; Champ, P.A.; Bond, C.A. 2011. Valuing the forest for the trees: Willingness to pay for white pine blister rust management. In: Keane, R.E., Tomback, D.F., Murray, M.P.; Smith, C.M. (eds.), The Future of High-Elevation, Five-Needle White Pines in Western North America: Proceedings of the High Five Symposium. 28-30 June 2010, Missoula, MT. Proceedings RMRS-P-63. Fort Collins, CO: U.S. Department of Agriculture, Forest Service, Rocky Mountain Research Station. pp. 226-234.

Mendelsohn, R.; Markstrom, D. 1988. The use of travel cost and hedonic methods in assessing environmental benefits. In: Peterson, G.L.; Driver, B.L.; Gregory, R., eds. Amenity resource valuation: integrating economics with other disciplines. State College, PA: Venture Publishing. pp. 159-166.

Michalson, E.L. 1975. Economic impact of mountain pine beetle on outdoor recreation. Southern Journal of Agricultural Economics. 7(2): 43-50.

Michigan Department of Natural Resources and Environment [Homepage of DNR – Balsam Woolly Adelgid], [Online]. (2001). Available: http://www.michigan.gov/dnr/0,1607,7-153-10370_12145_12204-33877--,00.html [September 7, 2010].

Millar, C.I.; Neilson, R.; Bachelet, D.; Drapek, R.; Lenihan, J. 2006. Climate Change at Multiple Scales. In: Salwasser, H.; Cloughesy, M., eds. Forests, carbon, and climate change: a synthesis of science findings. Portland, OR: Oregon Forest Resource Institute. Pp. 31-62.

Millar, C.I.; Stephenson, N.L.; Stephens, S.L. 2007. Climate change and forests of the future: Managing in the face of uncertainty. Ecological Applications. 17(8): 2145-2151.

Miller, J.D.; Lindsay, B.E. 1993. Willingness to pay for a state gypsy moth control program in New Hampshire: a contingent valuation case study. Journal of Economic Entomology. 86(3): 828-837.

Mitchell, R.C.; Carson, R.T. 1989. Using surveys to value public goods: the contingent valuation method. Washington, DC: Resources for the Future. 463 p.

Moeller, G.H.; Marler, R.L.; McCay, R.E.; White, W.B. 1977. Economic analysis of the gypsy moth problem in the northeast. III: Impacts on homeowners and managers of recreation areas. Res. Pap. NE-360. Upper Darby, PA: U.S. Department of Agriculture, Forest Service, Northeastern Forest Experiment Station. 9 p.

Moore, J.A.; Hatch, C.R. 1981. A simulation approach for predicting the effect of Douglas-fir tussock moth defoliation on juvenile tree growth and stand dynamics. Forest Science. 27(4): 685-700.

Payne, B.R.; White, W.B.; McCay, R.E.; McNichols, R.R. 1973. Economic analysis of the gypsy moth problem in the northeast. II: Applied to residential property. Res. Pap. NE-285. Upper Darby, PA: U.S. Department of Agriculture, Forest Service, Northeastern Forest Experiment Station. 6 p.

Pendleton, L.; Atiyah, P.; Moorthy, A. 2007. Is the non-market literature adequate to support costal and marine management? Ocean & Coastal Management. 50(5-6): 363-378.

Price, J.I.; McCollum, D.W.; Berrens, R.P. 2010. Insect infestation and residential property values: A hedonic analysis of the mountain pine beetle epidemic. Forest Policy and Economics. 12(6): 415-422.

Romme, W.H.; Knight, D.H.; Yavitt, J.B. 1986. Mountain pine beetle outbreaks in the Rocky Mountains: Regulators of primary productivity? The American Naturalist. 127(4): 484-494.

Rosenberger, R.S. 2006. Social and economic issues of global climate change in the western United States. In: Joyce, L.; Haynes, R.; White, R.; Barbour, R.J., comps. Bringing climate change into natural resource management: proceedings. Gen. Tech. Rep. PNW-GTR-706. Portland, OR: U.S. Department of Agriculture, Forest Service, Pacific Northwest Research Station. 150 p.

Rosenberger, R.S.; English, D.B.K. 2005. Impacts of wilderness on local economic development. In: Cordell, H.K., Bergstrom, J.C., and Bowker, J.M. (eds.), The Multiple Values of Wilderness. State College, PA: Venture Publishing, Inc. pp. 181-204.

Rosenberger, R.S.; Loomis, J.B. 2003. Benefit transfer. In: Champ, P.A.; Boyle, K.J.; Brown, T.C., eds. A primer on nonmarket valuation. Dordrecht, The Netherlands: Kluwer. pp. 445-482.

Rosenberger, R.S.; Phipps, T.T. 2007. Correspondence and convergence in benefit transfer accuracy: Meta-analytic review of the literature. In: Navrud, S.; Ready, R., eds. Environmental value transfer: Issues and methods. Dordrecht, The Netherlands: Springer. pp. 23-43.

Rosenberger, R.S.; Smith, E.L. 1997. Nonmarket economic impacts of forest insect pests: a literature review. Gen. Tech. Rep. PSW-GTR-164. Albany, CA: U.S. Department of Agriculture, Forest Service, Pacific Southwest Research Station. 38 p.

Rosenthal, D.H.; Loomis, J.B.; Peterson, G.L. 1984. The travel cost model: concepts and applications. Gen. Tech. Rep. RM-109. Fort Collins, CO: U.S. Department of Agriculture, Forest Service, Rocky Mountain Forest and Range Experiment Station. 10 p.

Rosner, H. 2009. A slow-moving disaster. High Country News. Natl. ed. (1): 8-9, 24.

Thompson, R.; Hanna, R.; Noel, J.; Piirto, D. 1999. Valuation of tree aesthetics on small urban-interface properties. Journal of Arborculture. 25(5): 225-234.

U.S. Water Resources Council. 1983. Economic and environmental principles and guidelines for water related land implementation studies. Washington, DC: U.S. Water Resources Council. 137 p.

Walsh, R.G.; Bjonback, R.D.; Aiken, R.A.; Rosenthal, D.H. 1990. Estimating the public benefits of protecting forest quality. Journal of Environmental Management. 30(2): 175-189.

Walsh, R.G.; Keleta, G.; Olienyk, J.P. 1981a. Value of trees to residential property owners with mountain pine beetle and spruce budworm damage in the Colorado Front Range. Fort Collins, CO: Colorado State University for U.S. Department of Agriculture, Forest Service (Cooperative Agreement No. 16-950-CA; and Contract No. 53-82X9-9-180). 116 p.

Walsh, R.G.; Keleta, G.; Olienyk, J.P.; Waples, E.O. 1981b. Appraised market value of trees on residential property with mountain pine beetle and spruce budworm damage in the Colorado Front Range. Fort Collins, CO: Colorado State University for U.S. Department of Agriculture, Forest Service (Cooperative Agreement 16-950-CA). 150 p.

Walsh, R.G.; Olienyk, J.P. 1981. Recreation demand effects of mountain pine beetle damage to the quality of forest recreation resources in the Colorado Front Range. Fort Collins, CO: Colorado State University for U.S. Department of Agriculture, Forest Service (Contract No. 53-82X9-9-180). 87 p.

Walsh, R.G.; Ward, F.A.; Olienyk, J.P. 1989. Recreational demand for trees in national forests. Journal of Environmental Management. 28(3): 255-268.

Werner, R.A.; Holsten, E.H. 1985. Factors influencing generation times of spruce beetles in Alaska. Canadian Journal of Forest Research. 15(2): 438-443.

Wickman, B.E.; Renton, D.A. 1975. Evaluating damage caused to a campground by Douglas-fir tussock moth. Res. Note PNW-257. Portland, OR: U.S. Department of Agriculture, Forest Service, Pacific Northwest Forest and Range Experiment Station. 5 p.

Wickman, B.E.; Mason, R.A.; Trostle, G.C. 1998. Douglas-fir Tussock Moth. [Online]. Available: http://www.na.fs.fed.us/spfo/pubs/fidls/tussock/fidl-tuss.htm [2011, June 29].

Appendix A: Economic Nonmarket Valuation Methods

The supply of ecosystem goods and services has declined, while the demand for them has grown over the past half-century. These trends are largely driven by land use and land cover changes to meet the needs of an increasing population, and an emergence of a greater understanding and appreciation of ecosystem goods and services. Concurrent with this trend has been the development of nonmarket valuation techniques for quantifying the benefits of ecosystem goods and services. The economic value of an environmental resource is defined as the most an individual would be willing to pay in order to obtain that good, service, or state of the world (King and Mazzotta 2000). By aggregating individual values, social worth can be calculated. A person's willingness to pay, usually measured in dollars, reveals how much a person is willing to give up in other goods and services in order to gain a specific change in quantity or quality of another good or service (King and Mazzotta 2000; Rosenberger and Smith 1997). Total willingness to pay is the amount actually paid for the good or service plus the additional amount a person would be willing to pay for it (also known as consumer surplus). Net willingness to pay is equal to the total willingness to pay less the amount actually paid for the good or service. Marginal willingness to pay is the amount a person will pay for the last unit purchased of a good or service. All of the economic nonmarket valuation methods used in the studies presented in this document—travel cost, hedonic pricing, contingent valuation and contingent choice methods—are derived from consumer theory (Champ and others 2003).

Travel Cost Method

The travel cost method is a valuation approach that uses variations in the costs of travel and other expenditures as implicit prices of destination sites in order to estimate the demand for the site. This method is typically used to estimate the onsite benefits of particular sites. It indirectly estimates the economic worth of a nonmarket resource by first defining the demand for the resource as a function of travel costs. Distance is an important factor in defining the demand curve. Travel costs are, in part, the variable costs of the trip. Therefore, the farther individuals are from the site, the more they pay in travel costs to the site, and in general, the fewer trips they take to the site. Different resource conditions (i.e., forest health levels) will result in shifts in the demand function for the resource. Economic worth of the resource change is measured as consumer surplus, which is equivalent to net willingness to pay. Consumer surplus is the amount of benefit gained from the purchase of a good or service beyond what was actually paid (entrance fees, travel costs) and is usually measured in monetary units.

Changes in forest health caused by pests are likely to shift the demand for recreation at the site, thus affecting the overall benefits derived from the experience. Two basic versions of the travel cost model are the individual and the zonal method. The individual travel cost method defines the number of visits to a site for a given time period for an individual. The zonal travel cost method creates a set of zones from which visitors originate, and then defines the visitation rate from the different zones (where visitation rate is the number of visits from a zone divided by the population of that zone). For more detailed discussions concerning the travel cost method, see Rosenthal and others (1984), Loomis and Walsh (1997), Champ and others (2003), and King and Mazzotta (2000).

Hedonic Pricing Method

The hedonic pricing method is another indirect estimation technique used in the economic valuation of nonmarket resources. This method estimates economic worth as a hedonic price, or the worth of an attribute as it is associated with the overall market price of a multi-attribute good such as a house or parcel of land. It is based on the variation in selling prices in an actual market and assumes that these variations are correlated with the presence of differing levels of specific attributes. For example, the price of a house may be based on the number of rooms, location, nearness to open space or public facilities, proximity of forests, and the like.

Hedonic pricing assumes that consumers choose market goods based on certain identifiable characteristics or attributes. It views any good as a bundle of attributes, each attribute contributing to the overall market worth of the good. Hedonic pricing statistically determines the marginal contribution of an attribute to the overall benefit of owning a good having the attribute by means of a two-stage process. Usually it relies on a sufficient number of sales transactions in order to allocate worth to specific attributes. For instance, assume there are two residences identical in all attributes except one: forest quality. If the residence with a preferred forest characteristic costs $X more than the other residence, then the forest attribute can be allocated a worth of $X. This is the implicit price paid for the attribute in the first stage of the analysis. In the second stage, the implicit price, along with other relevant information (such as the quality of the surrounding forest and socioeconomic variables) is used to estimate the demand for the attribute. Economic worth of the attribute is calculated as the area under the demand curve and above the implicit price line. The impact of pest-caused changes in forest health would be the reduced benefit derived from owning a good with forest health-dependent attributes. For more detailed discussions concerning the hedonic pricing method, see Freeman (1979), Brookshire and others (1982), Mendelsohn and Markstrom (1988), and Champ and others (2003).

Contingent Valuation Method

The contingent valuation method is a direct estimation method based on intended or stated behavior. This method asks individuals for their values (usually in monetary units) for defined changes in the quantity or quality of a good or service. Contingent valuation directly estimates economic worth as willingness to pay or compensation due by surveying or interviewing individuals. Contingent valuation constructs a hypothetical market in which the quantity and/or quality of a nonmarket resource is varied or changed, for example, forest conditions. Thus, the individual bids for the resource in a hypothetical market is contingent on the changes in the resource. The contingent valuation method assumes the individual can rationally express the economic worth of a nonmarket resource in this manner and that the individual's expressions are accurately elicited by the hypothetical situation.

Some concerns about the empirical application of this method have been raised, and procedural methods have been published to make its application more consistent (Arrow and others 1993; U.S. Water Resources Council 1983). Some of the concerns raised include many potential sources of bias in the results, such as interest, interviewer, and strategic biases. Another

concern is the proper amount of information to be included in the survey and the best way to convey this information. The choice of the proper payment vehicle (taxes, donations, and the like) has also led to numerous arguments. Yet another concern is what the proper elicitation method should be. Some of the elicitation methods used include payment card, open-ended, dichotomous or discrete choice, and iterative bidding methods. Many concerns about these issues can be overcome with proper survey design. For more detailed discussions concerning the contingent valuation method, see Mitchell and Carson (1989), Loomis and Walsh (1997), and Champ and others (2003).

Contingent Choice Method

The contingent choice method, like contingent valuation, is used to estimate the economic value of ecosystems and environmental services through stated preference. This method can estimate use and passive-use values (including option, existence, and bequest values) by analyzing the choices people make when asked about a hypothetical scenario. The hypothetical scenario is carefully constructed by the researchers and the questions asked elicit choices or tradeoffs between different amenities, such as aesthetic quality, recreation opportunities, and environmental quality, described in terms of its characteristics or level of attributes.

Sometimes referred to as conjoint analysis, contingent choice was developed to measure preferences for different characteristics or attributes of a multi-attribute choice. This method assumes that choices are made up of many attributes, including price, and that an individual can rank or choose between varying levels or amounts of each attribute. If price is included as one of the attributes of the good, then willingness to pay for changes in the attributes can be calculated as the tradeoff between a resource level or condition and income (i.e., money). Because the contingent choice method can value actions as a whole, as well as the individual parts of an action, it is very compatible with the policy decision-making process, especially when the decision being made impacts a natural resource or ecosystem service. For more information on the contingent choice method, see Haefele and Loomis (2001), King and Mazzotta (2000), and Champ and others (2003).

Appendix B: Extended Summaries of Studies

This appendix includes extended summaries of reviewed studies that are either excerpted from Rosenberger and Smith (1997) (studies #1-15, which are direct quotes) or are new summaries of recently added studies (#16-22). NOTE: The economic values reported in the summaries of the studies are all given in the dollar value from the year of publication. Some of the estimates in a particular study are based on the worth of the property at the time of the study; therefore, it may not be correct to assume that property values have increased at the same rate as the price index.

1. Payne and others 1973. Residential property in the Northeast and the Gypsy Moth (Appendix Table 1)

Payne and others (1973) estimate the gypsy moth's impact on residential property value in the northeastern United States. Gypsy moths are most destructive in residential areas where the worth of trees for conversion to wood products is greatly exceeded by the amenity benefits produced by the trees. Federal, state, and local agencies need estimates for the impacts of gypsy moths on the residential property in order to improve decisions on controlling gypsy moths. The authors present a method for estimating the losses in residential property values due to gypsy moths.

This study uses the hedonic pricing method to determine the contributed worth of trees to residential property where the trees are not normally separable from the land parcel itself. The amenity benefits of trees for property owners include shade, microclimatic effects, and aesthetics, and are reflected in the market price of the property. A gypsy moth attack, through repeated defoliation, can kill the tree and completely eliminate the production of its amenity benefits until a different tree has matured to the point of total replacement for the lost tree. The authors test the Amherst model by applying it in a similar environment: Stroudsburg, Pennsylvania. The data include the fair market value of the parcel, the size of the parcel, the number of trees [≥6 inches diameter at breast height (dbh)], and an estimate of the level of tree mortality associated with a given level of insect infestation.

In the pre-attack scenario (with a gypsy moth control program of 100 percent effectiveness), it is estimated that the contributed worth of trees to residential property

is $7,767/acre with 29 trees/acre; or about $270/tree. Incremental per tree benefits represent diminishing returns per tree. Beyond this level of 29 trees/acre, each incremental tree adds less to the property value (up to the observed maximum of 50 trees per acre in the sample). The average arc elasticity of demand for trees on residential property is 0.24, meaning that a 0.24-percent reduction in the contributed worth of trees to residential property results from a 1-percent decrease in the number of trees 6 inches dbh.

In the post-attack period (without a gypsy moth control program), the loss in property value is equal to the predicted number of trees killed times the estimated per-tree worth, according to this study's estimates. The net worth of the change in property value after attack by gypsy moths is the difference between the pre-attack and post-attack estimates. Therefore, a gypsy moth attack with a 15-percent mortality rate (29 trees per acre to 24.65 trees per acre) results in a loss of approximately $1,175/acre. The estimated worth is only for the benefits accruing to the onsite property owners. Benefits not measured are the offsite values accruing to the general population (neighbors, passers-by). Impacts not accounted for include non-mortality effects like discoloration and defoliation of trees, and the unpleasant effects of the presence of moths (caterpillars, feces, cocoons, allergic reactions to moths). Other losses associated with pest infestations include the removal and replacement costs of dead trees, and the costs of control. The model used is applicable only to areas similar to the two study areas.

2. Wickman and Renton 1975. Esthetics at the Stowe Reservoir Campground in California and the Douglas-fir Tussock Moth (Appendix Table 2)

Wickman and Renton (1975) estimate the total loss of recreation benefits of camping at a Forest Service campground at Stowe Reservoir in the Warner Mountains, Modoc National Forest, California, caused by the Douglas-fir tussock moth infestation. The moth can cause heavy defoliation of white fir, resulting in tree mortality and top-kill. This damage can have a significant impact on the recreation benefits at specific sites by reducing the amount of shade, aesthetic quality, and privacy screening. The authors estimate the insect damages by adding the actual costs of clean-up to a calculated unit worth of trees as aesthetic value.

The aesthetic worth component of trees in the campground is equivalent to the total replacement costs of the campground. With 46 trees on each of eight camp units, the aesthetic value of the campground based on estimated replacement costs is $20,610 (about $2,576 per camp unit or $56 per tree). The actual cost of cleanup (felling, removal, and topping of affected trees) is estimated to be $324 for the campground with 25 trees killed due to insect infestation. The total loss in campground worth if 25 trees are killed is estimated to be about $1,725 (total aesthetic damage plus cleanup costs), or about $216 per camp unit.

The method employed in this study may be sufficient to estimate nonmarket benefits if the economic evaluation is concerned with cost-minimization or cost savings approaches on the public good supply side. However, on the demand side, the manager may be more interested in the benefits generated by the resource (in this study, the camp unit with pre-infestation characteristics). The actual and replacement cost valuation approach may not adequately estimate nonmarket benefits (Loomis 1993). Total replacement costs of a camp unit may not be a good proxy for the total worth of the unit if it does not account for potentially high benefits generated by the unit (aesthetic value). The replacement costs also may not be feasible for a 60- to 70-year-old stand of white fir. If anything, the loss of nonmarket benefits will continue to accrue until replacement trees grow to sufficient size, mitigating the damage caused by the insects.

3. Michalson 1975. Recreationists in the Targhee National Forest and the Mountain Pine Beetle (Appendix Table 3)

Michalson (1975) estimates the impact of the mountain pine beetle on the benefits of recreation in the Targhee National Forest in eastern Idaho. Mountain pine beetle infestation in the predominantly lodgepole pine forested area affects recreation by killing trees, thus reducing aesthetics in the short run. This study estimates the impact on recreation of the increased presence of dead trees caused by a mountain pine beetle infestation. The purpose of this study is to estimate recreation losses, allowing the decision maker to determine the amount of investment needed, if a control program is to be implemented.

The author estimates the losses to recreation in the study area by means of the travel cost method, using expenditure and visitor-day loss calculations. He calculates the impact of the beetles by estimating the consumer surplus and expenditure per visitor day, and the number of visitor days for campgrounds with >30-percent infestation and for campgrounds with <30-percent infestation. The study's hypothesis is that the difference between these estimates equals the amount of recreation loss, or economic impact, associated with the mountain pine beetle damage. Although, as discussed earlier, expenditure is not a good measure of economic loss, it is a helpful measure if a cost-minimization approach is chosen.

Michalson (1975) estimates that a scenario without mountain pine beetle infestation generated $30.49 in average consumer surplus per visitor day, per person, with each visitor spending on average $5.28 per visitor day and staying approximately 3.3 visitor days per person per visit. The estimate for a scenario with mountain pine beetle infestation is $26.40 in average consumer surplus per visitor day, per person, with each visitor spending on average $4.85 per visitor day and staying approximately 2.1 visitor days per visit. The difference between the estimates for the scenario without infestation and those for the scenario with infestation is the economic impact of a mountain pine beetle infestation. These impact estimates, on average, are a loss of $4.09 in consumer surplus per visitor day per person, a reduction in the amount spent per visitor day per person of $0.43, and a reduction in length of stay per visit of 1.2 visitor days per person. Based on an estimated 124,783 visitor days per year (in the early 1970's), there is an estimated loss of $510,362 in consumer surplus per year in the Targhee National Forest, and a loss in local expenditures of $53,657 per year with the level of damage at the time of the study.

4. Moeller and others 1977. Homeowners and Recreation Area Managers in the Northeast and the Gypsy Moth (Appendix Table 4)

Moeller and others (1977) estimate the gypsy moth's impact on homeowners and managers of recreation areas as control costs, financial losses, and lost recreational use. The authors identify five ownership classes: (1) homeowners using public control methods, (2) homeowners using commercial control methods, (3) commercial campgrounds, (4) quasi-public recreation areas, and (5) public recreation areas. Federal, state, and local agencies need estimates of the gypsy moth's impact on the economic and ownership objectives of the different ownership classes. Decision makers can use the information in the design, implementation, and coordination of gypsy moth control programs.

Gypsy moth infestations can affect specific objectives of the owner/manager of a property. The authors interviewed 540 homeowners and 170 managers of recreation areas in Pennsylvania and New York in 1973. The authors identify four objectives for each ownership category that can be affected by gypsy moths. Homeowner objectives, excluding the need for a place of residence, include (1) the enjoyment of natural beauty, (2) backyard recreation, (3) property value, and (4) the maximization of recreation use. The study identifies the following management objectives for managers of recreation areas: (1) the maximization of recreation revenue, (2) the maximization of property value, (3) the maximization of the enjoyment of natural beauty, and (4) the maximization of recreation use. The authors identify four possible effects gypsy moth infestations can have on the ownership objectives: (1) nuisance (presence of insects, feces, and egg masses), (2) defoliation of trees and shrubs (reducing the shade and aesthetic quality of the property), (3) mortality (the killing of trees by the moths), and (4) other (allergic reactions to the moths and the like).

Calculations of important indices provide a measure of the relative importance of the impact of moths on a specific ownership objective. The study found that the nuisance effect affected the enjoyment of natural beauty and backyard recreation the most for homeowners, regardless of whether public or commercial control methods were employed. The nuisance effect of a gypsy moth infestation also affects the enjoyment of natural beauty in recreation areas the most, in the opinion of managers of commercial campgrounds and managers of quasi-public recreation areas (such as land owned and operated by the Boy Scouts organization). Managers of public recreation areas believed that the maximization of recreation revenue and enjoyment of natural beauty are affected most by the nuisance effect associated with a gypsy moth infestation. The only other effect that shows significant impacts on specific management objectives is defoliation.

The authors also report financial information on the control costs of a gypsy moth infestation, as well as financial losses and economic information on recreation losses for the different ownership classes attributable to the infestation. Control costs represent an economic measure of the importance owners place on goods and services and include what the owners actually spend on gypsy moth control programs, including the cost of equipment, materials, services,

and the owners' own labor input. This study assigns labor a worth of $2 (in 1977 dollars) per hour in moth control. Homeowners who participate in a public control program spend on average $102 per year, whereas those who participate in commercial control methods spend on average $240 per year. Managers of recreation areas spend $441 on average for commercial campgrounds and $722 per year for quasi-public recreation areas in control costs. No data were collected for managers of public recreation areas for either control costs or financial losses.

Financial losses are another measure of the impacts of gypsy moths on property through the capital cost reduction in property value and the increase in maintenance costs and revenue losses. Homeowners who participate in public control programs report $125 in financial losses per year on average, whereas those who participate in commercial control programs report losses of $479 per year on average. Managers of recreation areas report financial losses of $249 per year for the average commercial campground operation, and $996 per year for the average quasi-public recreation area.

Effects of gypsy moth infestations can alter the recreation benefits derived from private and public property. The economic effects of a moth infestation on recreation are not included in the above estimates. Recreation use losses in the different ownership classes are estimated as the number of person-days of recreation use lost per year. Homeowners with public controls report a loss of 108 person-days in recreation use per year on average, whereas homeowners with commercial controls report an average recreation use loss of 133 person-days per year. Managers of recreation areas report larger impacts on recreation use than homeowners. Managers of commercial campgrounds report an average of 161 lost person-days of recreation use per year. Managers of quasi-public campgrounds report an average recreation use loss of 240 person-days per year. Managers of public recreation areas report an average loss of 36,660 person-days per year.

This study ranks the relative effects of a gypsy moth infestation on the different ownership/management objectives. It estimates financial as well as economic measures of gypsy moth impacts. It measures the economic losses of recreation use of property in person-days, which can be converted to a commensurable dollar metric if the dollar worth of a person-day in recreation use is known. This is the first study to address the direct impacts of the presence of gypsy moths through the nuisance effect. Most studies estimate the secondary effects of a pest infestation, such as tree mortality, defoliation, and discoloration, which result in reduced recreation quality, aesthetic or scenic quality, property value, and other amenity benefits produced by trees or forests, ignoring the direct effects the presence of insects may have on benefits.

5. Leuschner and Young 1978. Recreation at East Texas Reservoir Campsites and the Southern Pine Beetle (Appendix Table 5)

Leuschner and Young (1978) estimate the southern pine beetle's impact on recreation use of reservoir campsites in east Texas, where beetles kill patches of trees near the reservoirs. This affects many different forest products, including the recreation benefits of the surrounding campgrounds. To deal with these effects, forest managers need to be able to quantitatively evaluate pest control programs. One method of quantitative analysis is through the use of benefit-cost analysis. In order to use benefit cost analysis, all relevant values must be measured and included. Recreation benefit is negatively affected because the beetle-killed trees reduce the shade available and the number of living trees in campsites. Normally, two reactions will be forthcoming following a beetle infestation: (1) the recreationists will continue to use the site but at a reduced enjoyment level; or (2) the recreationists will substitute unaffected sites for their recreation outings. The authors report that they do not attempt to estimate the first reaction because they lack a state-of-the-art method for estimating this form of a recreation impact.

The authors estimate the demand functions for two types of campgrounds on the basis of secondary data collected by the managing agencies. The two managing agencies are the USDA Forest Service (FS) and the U.S. Army Corps of Engineers (COE). Seven FS and 12 COE campgrounds are included in the study. The impact indicator variable is the percentage of pine crown cover, estimated from aerial photos taken in 1969 (black and white photos) and in 1970 (color photos). Secondary data on recreation use collected by the managing agencies on site are used to estimate the recreation demand for campgrounds (based, in part, on the percentage of pine crown cover) through the use of the zonal travel cost method. The unaffected recreation benefit is estimated as the area under the derived demand curve. By changing the percentage of pine crown cover in the derived demand functions for recreation at different campgrounds to simulate beetle infestations, the authors were able to predict the impact of a beetle infestation on the recreation benefit at the site. They estimate total recreation benefit for the campgrounds included in the study as $12,350,800 to $19,889,600, depending on assumptions about the worth of travel time to the recreation site. The smaller amount is when travel time costs are not included, and the larger amount is when a positive travel time cost is allocated on the basis of a methodology current at the time of the study (Cesario 1976).

The recreation damages incurred from a beetle infestation are calculated for each of the campgrounds in the study. Aggregate impacts cannot be calculated because (1) unaffected campgrounds are substituted, by campers, in place of infested campgrounds, (2) some of the sites are not included in the simulated pest infestation, (3) the probability is low that all sites would be affected identically, and (4) there is no estimate for the decreased recreation benefit for those who continue to use the affected site. However, a rough aggregate estimate can be calculated assuming all sites are equally affected and no substitution occurs. Recreation benefits at FS campgrounds, with a 10-percent reduction in pine crown cover, are negatively impacted by $1.12 per visit. With a 30-percent reduction in pine crown cover, recreation benefits decrease by $3.37 per visit. This results in an estimated recreation loss at FS campgrounds of approximately $287,400 with a 10-percent reduction in pine crown cover and $862,300 with a 30-percent reduction. Recreation benefit at COE campgrounds with a 10-percent reduction in pine crown cover is negatively affected by $0.82 per visit. With a 30-percent reduction the decrease in benefit is $2.44 per visit. This results in an estimated loss at COE campgrounds of approximately $1,045,000 with a 10-percent reduction in pine crown cover and $3,113,000 for a 30-percent reduction. A total loss of $1,332,400 in recreation benefit at east Texas reservoirs is estimated with a 10-percent reduction in pine crown cover. Estimated total loss with a 30-percent reduction is $3,975,300.

The authors also investigate the effect of substitution on the recreation loss of a simulated pest infestation at two campgrounds. Site-specific damage estimates are reduced by 85 to 90 percent when unaffected sites are substituted for the attacked site. However, the larger the affected area, the smaller this reduction is, because few or no unaffected substitute sites are available. The damages incurred from a beetle infestation are short-term, but multi-period. These damages may eventually be mitigated through the natural regeneration of forest quality through regrowth, but during the replacement and regrowth periods, some positive level of recreation losses are realized.

6. Walsh and Olienyk 1981. Recreation in the Colorado Front Range and the Mountain Pine Beetle (Appendix Table 6)

Walsh and Olienyk (1981) estimate the impacts of the mountain pine beetle on recreation demand in the Front Range of the Colorado Rocky Mountains. Recreation activities include developed camping, semi-developed camping, backpacking, hiking, fishing, picnicking, and using off-road vehicles (ORVs). Mountain pine beetles attack and kill ponderosa pines, resulting in the short-run discoloration of needles and dead and down trees that detract from the perceived quality of the forests. The long-run effect of a beetle infestation is a reduction in the density of the forest. Both the very short-run and the short-run effects on forest quality affect the demand for recreation use of these forests. The results of the study contributed to the assessment of a USDA Forest Service forest insect and disease management program including the assessment of forest insect control programs, and of citizen participation in management decisions and cost-sharing programs.

A stratified random sample of 435 recreation users was interviewed onsite at six different forest recreation sites in 1980. Using contingent valuation (with an iterative bidding technique) and individual travel cost methods, the authors estimate the worth of these beetle impacts on recreation demand as willingness to pay (in dollars) and willingness to participate (in user days). The beetle contingent changes in forest quality (depicted through the use of color photos) investigated by the authors include the following indicators: (1) the number of trees 6 inches in dbh or more on the site, on adjacent land affecting the near view, and on distant land affecting the far view, (2) the size of trees, (3) the presence of visible beetle damage, (4) the presence of dead and down trees, (5) the distribution of trees over the area (the presence of treeless patches caused by a beetle infestation), and (6) the presence of large specimen trees.

The authors estimate the average arc elasticities for aggregate recreation demand (excluding ORV use) in a 1- to 15-percent decrease from the predicted level of the indicator variable with a mean number of 178 trees per acre. The elasticities show that a 1-percent decrease in the number of live trees per acre onsite results in a 0.28-percent decrease in recreation demand. A 1-percent decrease in the number of live trees on adjacent property affecting the near view results in a 0.25-percent decrease in the demand for recreation onsite. A 1-percent decrease in the number of live trees on distant land affecting the far view results in a 0.16-percent decrease in the demand for recreation onsite. Recreation demand decreases by 0.32 percent with a 1-percent decrease in the average size of the trees surviving a beetle infestation (calculated in the 3-inch to 12-inch dbh range). Recreation demand decreases by 2.30 percent with either a 1-percent increase in visible damage such as needle discoloration or a 1-percent increase in the presence of dead and down trees with slash on the ground. A 1-percent increase in treeless areas caused by a beetle infestation results in a 0.24-percent decrease in the demand for recreation onsite. Recreation demand decreases by 2.20 percent with a 1-percent decrease in the presence of large specimen trees bigger than 24 inches dbh (this is approximately two specimen trees) at the recreation site. Evidently, recreation demand is more sensitive to visible damage and the presence of large trees than it is to the other factors.

Walsh and Olienyk (1981) estimate the loss in user days of recreation use for the Colorado Rocky Mountains Front Range as the result of a 15-percent effect on the following forest attributes by a beetle infestation. Aggregate recreation demand (excluding ORV use) decreases by 370,000 user days per year with a 15-percent reduction in the number of trees onsite; by 330,000 user days per year with a 15-percent reduction in the number of trees on adjacent land; and by 207,000 user days per year with a 15-percent reduction in the number of trees on distant land. A 15-percent reduction in the average size of the trees on the recreation site reduces demand by 422,000 user days per year. Either a 15-percent increase in the percentage of visible beetle damage or a 15-percent increase in the amount of dead and down trees decreases recreation demand by 3,045,000 user days per year. If 15 percent of the recreation site is a contiguous treeless area as the result of beetle infestations, recreation demand for the site decreases by 317,000 user days per year. A 1-percent decrease in the number of large specimen trees on the recreation site decreases recreation demand by 192,700 user days per year. The authors also report the impacts of a beetle infestation on the demand for the individual recreation activities.

The authors also estimate the impact on consumer surplus per user day with a 15-percent reduction in the number of live trees 6 inches in dbh or more, per acre, by means of the contingent valuation method. They estimate a range of $1.50 to $1.70 per user day in lost consumer surplus, depending upon the initial number of trees per acre. The $1.50 is calculated for 178 trees per acre, and the $1.70 is calculated for 270 trees per acre. Using the travel cost method of recreation demand analysis, the authors also estimate the effect of a 15-percent decrease in the number of trees per acre on the number of trips to a recreation site in the Colorado Front Range per person per year, and on the consumer surplus per person per trip. The study found that a 15-percent decrease in the number of trees per acre results in 0.16 fewer trips per person per year and in a reduction in consumer surplus of $11.60 per person per year, or an average reduction of $1.75 in consumer surplus per person per trip.

To account for the natural recovery of a forest from an insect infestation, Walsh and Olienyk (1981) developed a regrowth model, which adjusts the losses over time to reflect the natural regenerative abilities of the forest. The losses estimated are per year and will continue to be realized for Colorado Front Range forests until replacement trees are of sufficient size and quality so that the original losses are completely offset. Benefits not included in the present study are the benefits of forest quality to the general public who may be willing to pay for the preservation of forest quality, for the option of future use of the forest, for the knowledge that forest quality exists and is protected, and for the satisfaction from the bequeathing of forests and forest quality to future generations. Other sources of value held by the general public include psychological and ecological values. The psychological and ecological benefits associated with forest health may be large enough to exceed the economic estimates listed in this study.

7. Walsh and others 1981a. Residential Property Owners in the Colorado Front Range and the Mountain Pine Beetle and Western Spruce Budworm (Appendix Table 7)

Walsh and others (1981a) estimate the impacts of the mountain pine beetle and the western spruce budworm on the contributed worth of trees to residential property in the Front Range of the Colorado Rocky Mountains. Trees provide shade, aesthetic quality, wildlife habitat, privacy, and other amenity benefits to the owners of property. Mountain pine beetles attack and either kill or cause visible damage to ponderosa pine trees. Western spruce budworms attack Douglas-fir trees, causing extensive visible damage, but rarely kill the trees. Insect infestations that kill or cause visible damage to trees near residential mountain properties have an effect on the satisfaction derived from owning and living on mountain property, which is reflected in the property value. The study was conducted to develop and apply a procedure for measuring the effect of mountain pine

beetle and western spruce budworm infestations on the worth of trees to owners of residential property in the Front Range of the Colorado Rocky Mountains. The results of the study contributed to the assessment of a USDA Forest Service forest insect and disease management program including the assessment of forest insect control programs, and of citizen participation in management decisions and cost-sharing programs.

A representative sample of 64 mountain homeowners was interviewed at five different study sites along the Front Range of the Colorado Rocky Mountains in 1980. Using the contingent valuation method (with an iterative bidding technique), Walsh and others (1981a) estimated the insect impacts on the contributed worth of trees to mountain residential property as the owners' willingness to pay for different levels of forest quality. The contingent changes in forest quality caused by insect infestations investigated by the authors and depicted in color photos include (1) the number of trees 6 inches in dbh or more on the residential property and on adjacent property in the near view, (2) the size of the trees on the property, (3) an expert expectation of a severe insect infestation, (4) the presence of visible tree damage, (5) the distribution of ponderosa pine and Douglas-fir on the property, (6) the distribution of trees over the property (the presence of treeless patches caused by an insect infestation), and (7) the presence of large specimen trees. The study divides mountain properties into improved lots (lots where a mountain home has been constructed) and unimproved lots (plotted lots where a subdivision has been filed but where no mountain homes have been constructed).

The authors estimate the average arc elasticities for the contributed worth of trees to mountain properties in a 1- to 15-percent range decrease from the predicted optimal level of the indicator variable with a mean number of 212 trees per acre. The elasticities show that a 1-percent decrease in the number of trees 6 inches in dbh or more on the property results in a 0.34-percent decrease in the contributed worth of the trees to improved lots and a 0.28-percent decrease for unimproved lots. A 1-percent decrease in the number of trees on property adjacent to improved lots affecting the near view decreases the worth of the improved lots by 0.20 percent. The worth of improved and unimproved lots decreases by 0.35 percent and 0.33 percent, respectively, with a 1-percent decrease in the average size of the trees surviving a beetle infestation on the lot (calculated in the 3-inch to 12-inch dbh range). The worth of improved lots decreases by 0.76 percent and the worth of unimproved lots decreases by 0.61 percent with a 1-percent increase in an expert expectation of insect damage to trees on the lot. A 1-percent increase in visible damage caused by an insect infestation results in a 2.27-percent decrease in improved lot worth and a 1.80-percent decrease for unimproved lots. A 1-percent increase in treeless areas caused by insects results in a 0.29-percent decrease in improved lot worth and a 0.40-percent decrease in unimproved lot worth. The worth of improved lots and unimproved lots increase by 0.02 percent and 0.04 percent, respectively, with a 1-percent change in the distribution of tree species from ponderosa pine to Douglas-fir. A 1-percent decrease in the presence of large specimen trees bigger than 24 inches dbh (approximately two trees per acre) decreases improved lot worth by 3.64 percent and unimproved lot worth by 2.61 percent. Property value is evidently more sensitive to visible damage and the presence of large trees.

Walsh and others (1981a) estimated the dollar equivalents in reduced mountain property value due to a 15-percent change in the forest quality indicator variables (calculated with an average of 212 trees per acre and mountain lots averaging 1 acre per lot). A 15-percent reduction in the number of trees on the lot decreases improved lot worth by $984 and unimproved lot worth by $578. Improved lot worth decreases by an additional $602 if the number of trees on adjacent property affecting the near view decreases by 15 percent. A 15-percent reduction in the average tree size results in a $1,228 reduction in the worth of improved lots and a $783 reduction in that of unimproved lots. With an increase of 15 percent in expected damage due to an insect infestation, based on expert opinion, improved lot worth decreases by $2,351 and unimproved lot worth decreases by $1,364. A $7,034 reduction in the worth of improved lots and a $4,045 reduction in that of unimproved lots results from a 15-percent increase in visible insect-caused damage to trees. With 15 percent of the lot treeless because of an insect infestation, the worth of improved lots and unimproved lots decrease by $907 and $896, respectively. With a 15-percent change in the distribution of tree species from ponderosa pine to Douglas-fir results in an increase in the worth of improved lots by $61 and that of unimproved lots by $84. The increasing percentage of Douglas-fir on the property leads to increased property value probably because Douglas-fir is scarce in the elevation range in which the study was conducted. A 1-percent decrease in the number of large specimen trees per lot decreases the worth of improved lots by $918 and that of unimproved lots by $449.

To account for the natural recovery of a forest from an insect infestation, Walsh and others (1981a) developed a regrowth model. The regrowth model adjusts the losses over time as a result of the natural regeneration abilities of the affected forest. The losses estimated are per year and continue to be realized for Colorado Front Range forests until replacement trees are of sufficient size and quality so that the original losses are completely offset. Benefits not included in this study are the benefits of forest quality to the general public who may be willing to pay for the preservation of forest quality, and for option, existence, and bequest values. Other sources of value held by the general public include the psychological and ecological values. The psychological and ecological benefits associated with forest health may be large enough to exceed the economic estimates listed in this study.

8. Walsh and others 1981b. Appraised Market Value of Trees on Residential Mountain Properties in the Colorado Front Range and the Mountain Pine Beetle and Western Spruce Budworm (Appendix Table 8)

Walsh and others (1981b) estimate the impact of the mountain pine beetle and western spruce budworm on the market value of improved and unimproved mountain properties in the Front Range of the Colorado Rocky Mountains. Trees provide shade, aesthetic quality, wildlife habitat, privacy, and other amenity benefits to the owners of property. Mountain pine beetles attack and either kill or cause visible damage to ponderosa pine trees. Western spruce budworms attack Douglas-fir trees, causing extensive visible damage, but rarely kill the trees. Insect infestations that kill or cause visible damage to trees near residential mountain properties have an impact on the satisfaction derived from owning and living on mountain property, which is reflected in the property value. Real estate appraisers allocate worth to all marketable attributes of a property in appraising the property's market value for potential or current owners. This study was conducted to develop and apply a procedure for measuring the effect of mountain pine beetle and western spruce budworm infestations on the contributed worth of trees to residential property in the Front Range of the Colorado Rocky Mountains. The results of the study contributed to the assessment of a USDA Forest Service forest insect and disease management program including the assessment of forest insect control programs, and of citizen participation in management decisions and cost-sharing programs.

A representative sample of 21 real estate appraisers of Front Range mountain property were interviewed in 1980. Using the contingent valuation method (with the iterative bidding technique), Walsh and others (1981b) estimate insect impacts on the contributed worth of trees to mountain residential property through the professional opinions of mountain property real estate appraisers for changes in different forest quality indicator variables. The contingent changes in forest quality caused by insect infestations investigated by the authors and depicted in color photos include (1) the number of trees 6 inches in dbh or more on the residential property and on adjacent property in the near view, (2) the size of the trees on the property, (3) an expert expectation of a severe insect infestation, (4) the presence of visible tree damage on the property, on adjacent property affecting the near view, and on distant property affecting the far view, (5) the distribution of ponderosa pine and Douglas-fir on the property, (6) the distribution of trees over the property (the presence of treeless patches caused by an insect infestation), and (7) the presence of large specimen trees. The study divides mountain properties into improved lots (lots where a mountain home has been constructed) and unimproved lots (plotted lots where a subdivision has been filed but where no mountain homes have been constructed).

The authors estimate the average arc elasticities for the contributed worth of trees to mountain properties in a 1- to 15-percent decrease from the optimal level of the indicator variable with a mean number of 106 trees per acre. The elasticities show that a 1-percent decrease in the number of trees 6 inches in dbh or more on the property results in a 0.14-percent decrease in the appraised contributed worth of the trees to improved lots and a 0.15-percent decrease in that of unimproved lots. A 1-percent decrease in the number of trees on property adjacent to improved lots affecting the near view decreases the worth of these lots by an additional 0.12 percent. The worth of improved and unimproved lots decreases by 0.53 percent and 0.54 percent, respectively, with a 1-percent decrease in the average size of the trees on the lots (calculated in the 3-inch to 12-inch dbh range). The worth of improved lots decreases by 0.94 percent and that of unimproved lots decreases by 0.92 percent with a 1-percent increase in the expert expectation of insect damage to trees on the lot. A 1-percent increase in visible damage caused by an insect infestation results in a 2.48-percent decrease in the worth of improved lots and a 2.06-percent decrease in that of unimproved lots. Improved property decreases in worth by 1.07 percent and unimproved property decreases by 1.85 percent with a 1-percent increase in the visible

damage caused by insects on adjacent property affecting the near view from the residential property. A 1-percent increase in visible damage affecting the far view decreases in worth by 0.16 percent for improved property and by 0.64 percent for unimproved property. A 1-percent increase in treeless areas, due to insect infestations, results in a 0.29-percent decrease in the worth of improved and unimproved lots. The worth of improved and unimproved lots increases by 0.05 percent and 0.06 percent, respectively, with a 1-percent change in the distribution of tree species from ponderosa pine to Douglas fir. A 1-percent decrease in the presence of large specimen trees bigger than 24 inches dbh (approximately two trees per acre) decreases the worth of improved lots by 1.94 percent and that of unimproved lots by 2.70 percent. Appraiser estimation of property value is most sensitive to visible damage and the presence of large trees.

Walsh and others (1981b) estimate the dollar equivalents in reduced mountain property value due to a 15-percent change in the forest quality indicator variables (calculated with an average of 106 trees per acre and mountain lots averaging 1 acre per lot). A 15-percent reduction in the number of trees on the lot decreases the worth of improved lots by $209 and that of unimproved lots by $201. The worth of improved lots decreases by an additional $241 if the number of trees on adjacent property affecting the near view decreases by 15 percent. With a 15-percent reduction in the average tree size, improved lot and unimproved lot values decline by $1,155 and $1,014, respectively. With an increase of 15 percent in the expectation of damage due to an insect infestation based on expert opinion, the worth of improved lots decreases by $1,810 and that of unimproved lots decreases by $1,492. A $3,902 reduction in the worth of improved lots and a $2,715 reduction in that of unimproved lots results from a 15-percent increase in visible damage to trees. The worth of improved lots decreases by an additional $1,688 for visible damage affecting the near view and by $259 because of visible damage affecting the far view with a 15-percent increase in visible damage on surrounding property. The worth of unimproved lots decreases by an additional $2,445 for visible damage affecting the near view and by $848 because of visible damage affecting the far view with a 15-percent increase in visible damage on surrounding property. With 15 percent of the lot treeless due to an insect infestation, improved lot and unimproved lot worth decreases by $493 and $476, respectively. A 15-percent change in the distribution of tree species from ponderosa pine to Douglas-fir results in an increase in the worth of improved and unimproved lots by $102. The increasing percentage of Douglas-fir on the property leads to increased property value probably because Douglas-fir is scarce in the elevation range in which the study was conducted. A 1-percent decrease in the number of large specimen trees per lot decreases the worth of improved lots by $251 and unimproved lots by $294.

To account for the natural recovery of a forest from an insect infestation, the authors developed a regrowth model. The regrowth model adjusts the losses over time as a result of the natural regeneration abilities of the affected forest. The losses estimated are per year and continue to be realized for Colorado Front Range forests until replacement trees are of sufficient size and quality that the original losses are completely offset. Benefits not included in this study are the benefits of forest quality to the general public who may be willing to pay for the preservation of forest quality, and for option, existence, and bequest values. Other values potentially held by the general public include those derived from psychological and ecological values. The psychological and ecological benefits associated with forest health may be large enough to exceed the economic estimates listed.

The two samples used by Walsh and others (1981a and 1981b) (residential property owners and real estate appraisers, respectively) can be compared directly. Residential property owners reported the contributed worth of forest quality to their perceived property value. Real estate appraisers gave their professional opinions concerning what they believed trees contribute to the worth of residential mountain property. The average arc elasticities reported show that real estate appraisers preferred the number of trees per acre less than property owners did. Owners of property with large specimen trees appreciate them more highly than real estate appraisers do, supporting the belief that specific trees may carry larger psychological benefits for the owner of the property. Real estate appraisers place greater emphasis on insect impacts to unimproved property than do property owners. This may be because unimproved property relies more heavily on the contribution of natural assets to its overall market value than does improved property. Improved property includes structures such as homes that comprise a large portion of its overall market value. Also, owners of improved property probably spend more of their time on their property than do owners of unimproved property.

9. Loomis and Walsh 1988. Recreation and Tree Stand Characteristics in the Colorado Front Range (Appendix Table 9)

Loomis and Walsh (1988) estimate the net economic benefits of recreation in the Colorado Rocky Mountains Front Range as a function of tree stand density and tree size. This study is included because the data set used for analysis is the Walsh and Olienyk (1981) data set. Effects of tree stand density and tree size on recreation are the result of perceived impacts from mountain pine beetles. Information on the effects of tree stand density and tree size on recreation use and benefit is important to the management of forests for stocking rates and tree growth. Different tree densities and average tree size in a forest stand can give rise to different recreation activities. This study investigates the recreation use and benefits of six recreation activities: camping, picnicking, backpacking, hiking, fishing, and use of off-road vehicles (ORVs). The results of the study can be used to derive the management implications of intensive forest management (e.g., planting affected areas instead of relying solely on natural regeneration, silvicultural practices like thinning).

A stratified random sample of 435 recreation users of six forest recreation sites composed of mixed-age ponderosa pine in the Front Range of Colorado was interviewed onsite in summer 1980. The contingent valuation method was used to estimate the recreationist's maximum net willingness to pay for different quantities of trees per acre and for changes in average tree size per acre, the changes being presented to the respondents through the use of color photos. The implied factor that caused the changes in tree stand density and average tree size is the mountain pine beetle. Therefore, the survey shows that changes in net willingness to pay are contingent on changes in tree stand density and average tree size, in this case being the result of insect infestations.

Recreation activities considered in the survey were found to be positively related to tree stand density and tree size except for the use of ORVs, which is negatively related to these factors. This means that 5 out of 6 of the activities have increased benefits with increased tree stand density and average tree size. The estimated annual recreation benefits as maximum willingness to pay per visitor, estimated at an average of 200 trees per acre, are $145 for camping, $169 for picnicking, $161 for backpacking, $302 for hiking,

$321 for fishing, and $97 for the use of ORVs. The annual recreation benefits per visitor as a function of tree size are estimated to be $28 when the average tree size per stand is 2.5 inches dbh, and $210 with an average tree size per stand of 10.5 inches dbh. Other estimates reported include recreation benefits per visitor day of $5 when average tree size is 4 inches dbh, and of $13 when average tree size is 13 inches dbh.

10. Walsh and others 1989. Recreation and the Demand for Trees in National Forests in the Colorado Front Range and the Mountain Pine Beetle (Appendix Table 10)

Walsh and others (1989) estimate and compare the average benefits per recreation trip in the Colorado Front Range as a function of the number of trees per acre 6 inches in dbh or more. This study is based on the original Walsh and Olienyk (1981) data set. The main purpose of this paper is to compare the benefits measured by means of the contingent valuation method as dollars willingness to pay for forest quality (the number of trees 6 inches in dbh or more) and benefits estimated by means of the travel cost method as consumer surplus. The study is included in this report because it includes three estimates of average benefits per recreation trip that are not considered in the Walsh and Olienyk (1981) study.

Recreation demand functions are derived using contingent valuation and travel cost methods. One measure of the average benefit per recreation trip is estimated by means of the contingent valuation method, and two measures are estimated by means of the travel cost method using two different econometric regression procedures (ordinary least squares and two-stage least squares). The results of the study support the hypothesis that the contingent valuation and travel cost methods produce comparable estimates. The estimates reported can be included in an economic assessment of a forest management alternative that incorporates market and nonmarket recreation use benefits.

A stratified random sample of 435 recreation users of six forest recreation sites composed of mixed-age ponderosa pine in the Front Range of Colorado was interviewed onsite in summer 1980. The contingent valuation method is used to derive the recreationist's demand for trees as an essential part of the recreation experience. A subsample of 220 recreationists (excluding off-road vehicle users and nonresidents) was selected for participation in the travel cost method of demand analysis. The travel cost method indirectly

derives the recreation demand for trees. Both demand functions are derived on the basis of the changes in the number of trees per acre as represented in color photos. After the recreation demand functions are derived, total benefits can be calculated as willingness to pay for the contingent valuation method and as consumer surplus for the travel cost method.

Walsh and others (1989) estimate the net average recreation benefits per user day from the contingent valuation method as $24. This and the following estimates are based on approximately 178 trees per acre and 2.7 days per trip. From the travel cost method of demand analysis, net average recreation benefits per user day, using the ordinary least squares regression technique, are estimated as $26. From the travel cost method, using the two-stage least squares regression technique, net average recreation benefits per user day are estimated as $20. Statistical tests show that the estimates based on all three methods are not statistically different.

11. Walsh and others 1990. Total Economic Nonmarket Worth of Forest Quality in Colorado National Forests and the Mountain Pine Beetle and Western Spruce Budworm (Appendix Table 11)

Walsh and others (1990) estimate the total economic nonmarket worth or public benefits of protecting forest quality in National Forests in Colorado. Total economic nonmarket worth accrues from recreation use, option, existence, and bequest values (Randall and Stoll 1983). Recreation use value is the benefit derived from the recreation experience and is restricted to onsite direct use of the resource. Option, existence, and bequest values are nonuse or passive-use benefits and can be derived either onsite or offsite. Option value is the satisfaction of knowing that a resource is protected for its potential use in the future. Existence value is the satisfaction of knowing that a resource is protected for its own sake. Bequest value is the satisfaction of knowing that a resource is protected for the potential use of others, including family and future generations. Measuring passive-use value of protecting forest quality is important because an economic assessment of a management alternative that includes only direct-use benefits would understate the true worth of the forest, possibly resulting in a socially inefficient outcome. The general population, including users and nonusers of forest resources, are affected by changes in forest quality and may be willing to pay to protect forest health.

Therefore, in deciding between management alternatives, the total economic nonmarket worth of forest resources must be incorporated in the decision process.

A random sample of 198 households in the Fort Collins and surrounding rural areas was interviewed in 1983. The sample was found to be socially and demographically representative of Colorado residents. The study uses the contingent valuation method (with an iterative bidding technique) to estimate the maximum net total benefits and recreation use benefits of protecting forest quality. The indicator variable for forest quality used is tree density measured as the number of trees 6 inches in dbh or more as depicted in color photos. The households surveyed state their maximum net willingness to pay for different forest stand densities as the result of mountain pine beetle and western spruce budworm infestations.

Walsh and others (1990) find that total average annual willingness to pay per Colorado household for the protection of forest quality is $52 (estimated with an average tree stand density of 150 trees per acre). Recreation-use benefit is 27.4 percent of total benefits, or $14 per household per year. Nonuse or passive-use benefit (option, existence, and bequest) make up 72.6 percent of the total, or $38 per household per year. Option and existence benefit is $11 per household per year each, and bequest benefit is $16 per household per year. The results show that nonuse or passive-use benefits are more than three and a half times greater than recreation-use benefits. Therefore, assessments of management alternatives that rely on direct onsite use value alone greatly understate the total benefits of a resource quality protection program and may result in inefficient resource allocations. These results are consistent with those collected by Brown (1993) who showed that existence and bequest value estimates derived through the contingent valuation method can be two to 10 times larger than direct onsite recreation-use value.

12. Jakus and Smith 1991. Private and Public Landscape Amenities in the Pennsylvania/Maryland Area and the Gypsy Moth (Appendix Table 12)

Jakus and Smith (1991) collected data on households' willingness to pay for aesthetic benefits that accrue solely to their own household versus benefits that accrue to the neighborhood in general from different gypsy moth control programs in the south central Pennsylvania/north central Maryland area. One program sprayed the residential (privately owned) areas only, while the public program sprayed both residential and common areas in the neighborhood. The author's study used the data collected to compare use and nonuse benefits associated with protection of landscape amenities. The hypothesis of the research is whether contingent behavior questions can be used for measuring use and nonuse values derived from an environmental resource (such as landscape aesthetics) that provides both private and public benefits. The contingent valuation method, with dichotomous choice elicitation, was used in a telephone-mail (informative brochure)-telephone survey of a 10-county area; 436 surveys were completed.

Respondents were asked to bid their maximum willingness to pay for each of two public gypsy moth control programs. The programs bid on include (1) spraying of a bacterial insecticide on residential properties only, and (2) spraying of a bacterial insecticide on residential and surrounding public areas (local parks and greenways). From the data, two linear and two nonlinear models are estimated, both with and without sample selection correction. A sample selection adjustment is introduced to account for nonparticipants in the second telephone interview stage. The estimated average household willingness to pay per year for the uncorrected linear model with a 25-percent reduction in defoliation ranges from $348 to $352 for the private program, and from $395 to $474 for the public program. The uncorrected nonlinear model for a 25-percent reduction in defoliation estimates average annual household willingness to pay as ranging from $464 to $534 for the private program, and $608 to $670 for the public program. When sample selection is corrected for, the linear specification for the private program results in an average annual household willingness to pay of $254 to $271, and for the public program of $314 to $344. The corrected nonlinear model estimates the average annual household willingness to pay as $376 to $420 for the private program, and $511 to $527 for the public program.

The authors conclude that individuals distinguish between private and public services provided by a gypsy moth control program through protecting landscape amenities. The results show an increase of 12 to 36 percent in average annual household willingness to pay for a public-scope program over a private scope program with the uncorrected models, and an increase of 16 to 36 percent for the sample selection corrected models. Therefore, individuals derive both use and nonuse benefits from environmental resources that exhibit private and public goods characteristics. This can be an important motivation for the private support of public programs.

13. Haefele and others 1992. Total Economic Nonmarket Worth of Forest Quality in the Southern Appalachian Mountains and the Balsam Woolly Adelgid (Appendix Table 13)

Haefele and others (1992) estimate the total economic nonmarket worth of protecting forest quality in the Southern Appalachian Mountains in North Carolina, Tennessee, and Virginia. The study decomposes total economic nonmarket worth into use and nonuse or passive-use value components. Nonuse value is further decomposed into bequest and existence values. Over the past few decades, two major impacts have affected the sustainability of the spruce-fir ecosystem in the Appalachians. The first is the decline in the number of Fraser fir trees in the area. The balsam woolly adelgid attacks the Fraser fir, resulting in high tree mortality rates. The other impact is the result of atmospheric deposition, such as acid rain, which is reducing the red spruce population and its regrowth potential. The results show that the general public is willing to pay for forest quality protection in the eastern United States, that individuals value forests for more than their own personal direct use, and that nonuse or passive-use value is greater than recreation-use value in the total economic nonmarket worth of protecting forest quality.

A random sample of 1,200 households was surveyed through the mail within a 500-mile radius of Asheville, North Carolina, in 1991. The contingent valuation method was used to estimate the maximum net benefits as willingness to pay for forest quality protection in the Southern Appalachian Mountains. The elicitation methods employed in the study were modified payment card and discrete choice, allowing for a comparison of the two methods for consistency of benefit estimates. The households were also asked to partition total willingness to pay into its component values: use, existence, and bequest values. The indicator variable is visual quality as depicted in color photos with changes in visual quality as the result of insect infestation or atmospheric deposition. The households surveyed state their maximum willingness to pay contingent on changes in forest quality for two areas: along roads and trails, comprising approximately one third of the total forest area, and for the whole forest area.

The pretest and focus group results for this study are presented by Holmes and others (1990). Haefele and others (1992) find that total average annual willingness to pay per household for forest quality along roads and trails ranges from $19 for the modified payment card version to $63 for the discrete choice version. The estimated total average annual willingness to pay per household for the whole forest ranges from $22 for the modified payment card version to $107 for the discrete choice version. The disparity between the two method estimates may be due to anchoring in the modified payment card approach, in which respondents are conditioned by the bid levels, providing them with valuation clues. Discrete choice may also exhibit anchoring along with upward rounding and the desire to provide "correct" answers by the respondents, thus inflating willingness to pay bids. Walsh and others (1989a) show that discrete choice models typically return larger willingness to pay estimates than modified payment card and open-ended question formats.

The allocation of total economic nonmarket worth to its components is very similar for the two elicitation methods. Total benefits estimated by means of the modified payment card method ($22) are allocated as 8 percent for use, 59 percent for existence, and 30 percent for bequest value. This results in the average annual willingness to pay, for the use of a spruce-fir forest at a given quality level, of $2; the existence benefit of the forest being $13; and the bequest benefit of the forest being $7. These benefits, based on the modified payment card method, are very similar to those estimated in the pretest (Holmes and others 1990). The allocation of total economic nonmarket worth via the discrete choice method ($107) is 13 percent for use, 56 percent for existence, and 31 percent for bequest value. This results in average annual willingness to pay for the use of a spruce-fir forest at a given quality level of $14; existence benefits being $60 and bequest benefit being $33. The results show that when nonuse or passive use values are included in the benefit estimate of forest quality protection, the total benefits are 7 to 12 times greater than recreation use value alone. These results support the evidence in Walsh and others (1990) and Brown (1993). The efficient allocation of forest resources depends on the estimation of total economic worth.

14. Miller and Lindsay 1993. Support for a Gypsy Moth Control Program in New Hampshire (Appendix Table 14)

Miller and Lindsay (1993) estimate the public support for a gypsy moth control program in New Hampshire through residents' willingness to pay for the program. In 1981, the gypsy moth population peaked, causing severe defoliation of 195,000 acres out of 2,000,000 infested acres in New Hampshire forests. At the time of the study, New Hampshire did not have a state gypsy moth program, leaving towns, cities, landowners, and homeowners to bear the costs of controlling gypsy moths. The costs include control methods, clean-up, and tree loss. Other costs include the psychological and social costs such as aesthetic degradation, recreation loss, and nuisance factors. Other impacts include wildlife habitat changes in tree browse and protective foliage.

This study uses the contingent valuation method, with dichotomous choice elicitation of willingness to pay. Miller and Lindsay surveyed 669 households from three towns: Bow, Conway, and Exeter. Bow represents the towns that experienced severe defoliation and had implemented a municipality-wide moth control program. Conway represents areas that experienced moderate to severe defoliation but did not adopt any central control program. Exeter represents towns that experienced no appreciable gypsy moth-caused damage. The results show that Bow residents' mean and median annual willingness to pay per household is $84 and $62, respectively. Conway residents' mean and median annual willingness to pay per household is $55 and $31, respectively. And Exeter residents' mean and median annual willingness to pay per household is $56 and $27, respectively. Aggregating the annual willingness to pay of the three towns results in mean and median annual willingness to pay per household of $70 and $43, respectively. On a per-acre basis, the aggregate mean and median annual willingness to pay per household is $16 and $10, respectively. This results in aggregate public support for a statewide gypsy moth control program of $13 million to $28 million mean annual willingness to pay, and $8 million to $17 million median annual willingness to pay. These results show strong support for a gypsy moth control program in New Hampshire.

15. Holmes and Kramer 1996. Total Economic Nonmarket and Existence Worth of Forest Quality in the Southern Appalachian Mountains and the Balsam Woolly Adelgid (Appendix Table 15)

Holmes and Kramer (1996) estimate the total economic nonmarket and existence worth of protecting forest quality in the Southern Appalachian Mountains in North Carolina, Tennessee, and Virginia. Total nonmarket economic worth is defined as the summation of use and existence values, where use value is the utility (or satisfaction) derived from active use of the resource and existence value is the utility derived from the resource for all other reasons other than active use. The researchers investigate the economic measures of forest health protection for the boreal montane forest ecosystem,

75 percent of which is contained in the Great Smoky Mountains National Park. Over the past few decades, mortality of the spruce and fir trees in this ecosystem has increased dramatically. This is generally attributable to the balsam woolly adelgid and air pollution.

The households sampled are within a 500-mile radius of Asheville, North Carolina, and were surveyed through a mail-out, mail-back questionnaire. This study elicited willingness to pay for the protection of the remaining healthy spruce-fir forests through the use of the contingent valuation method with a dichotomous choice elicitation procedure. Of the 210 surveys returned with usable dichotomous choice responses, 175 were for users of the ecosystem who stated total economic nonmarket worth, and 35 were for nonusers of the ecosystem who stated existence worth only. This allows for testing whether the existence component can be distinguished from total economic value. The indicator of forest health was visual quality as depicted in color photos with changes in quality resulting from insect infestation or atmospheric deposition. This and other data collected were used in a test on the convergent validity of two contingent valuation elicitation methods: dichotomous choice and modified payment card (Holmes and Kramer 1995).

Holmes and Kramer (1996) estimate the median annual willingness to pay as $36 for users (representing total economic nonmarket worth) and as $11 for nonusers (representing the existence component). They found the two estimates to be statistically different. The study draws three conclusions. First, the responses are consistent with the compositional approach to total economic worth, i.e., a component is less than the whole. Second, existence worth is a distinct and substantial component of total economic worth of forest health (existence comprised approximately 30 percent of the total). And third, the existence component is distinct from the use component, based on economic characteristics of direct-use and passive-use values.

16. Thompson and others 1999. Valuation of Tree Aesthetics on Small Urban-Interface Properties (Appendix Table 16)

Thompson and others (1999) developed a hedonic pricing model to identify how different aspects of forest condition contribute to the value of wildland-urban interface properties in the Lake Tahoe Basin. The Lake Tahoe Basin is located on the border between California and Nevada. The forests surrounding the basin are of two types. On the Nevada side of the lake the forests are populated with mixed and pure Jeffrey pine

(*Pinus jeffreyi*) stands. The California side consists mainly of the Sierra Nevada mixed-conifer type (i.e., California white fir (*Abies concolor*), ponderosa pine (*Pinus ponderosa*), sugar pine (*Pinus lambertiana*), incensecedar (*Libocedrus decurrens*), California black oak (*Quercus kelloggii*), and Douglas-fir (*Pseudotsuga menziesii*). Trees on or near residential properties can provide a wide range of benefits to homeowners including wildlife habitat, energy and water savings, pollution reduction, and aesthetic value. The authors hypothesized that traditional housing valuation characteristics, along with forest aesthetics characteristics, would account for a property's price. A second hypothesis asserted that tree size, number of trees per acre, condition, and species would influence forest property values. Additional variables were used to further examine the effect of forest pests and the degree of infestation in trees.

The characteristics of property transactions between 1989 and 1994 were collected for the California side of the Lake Tahoe Basin. For each property in the sample, tree and plant groupings were sampled for stand structure, stand composition and forest condition. The results of the model suggested that removing smaller trees from wildland-urban forests can have an immediate impact on average tree size and improve the view shed of and from the home. By removing dense and diseased trees, homeowners can reduce fire risk, enhance the aesthetic value of their home, and add value to their property. "If thinned trees were those most heavily infected then property values could be enhanced an additional 5% and as much as 30% on properties with many infected trees" (p. 229). The results of the hedonic generalized least square model revealed that trees infected by an invasive forest pest (insect) can reduce property value by as much as $26,390 dollars on average.

The authors concluded that although Lake Tahoe Basin represents a special case real estate market, the results obtained from the study are relevant. While stand density and overall forest health may be proxies for aesthetic qualities of a property, they can further enhance property values in other ways.

17. Haefele and Loomis 2001. Using the Conjoint Analysis Technique for the Estimation of Passive Use Values of Forest Health (Appendix Table 17)

Haefele and Loomis (2001) use the contingent choice technique to examine alternative management programs for three different forest pest situations in the United States. The first pest scenario is that of the gypsy moth, *Lymantria dispar*, in the northeastern United States. The gypsy moth, an invasive pest, has a large effect on ornamental trees on private property and in recreation areas. The second scenario dealt with the western spruce budworm, *Choristoneura occidentalis*, one of the most prominent tree defoliating insects in the West. This insect has a large effect on the yields of commercial timber and is native to most forests in the Northwestern United States. The third scenario involves the southern pine beetle. The pine beetle, a native insect, affects commercial timber production and wildland areas where control methods are limited.

The authors gathered data by presenting the three insect infestation scenarios in a questionnaire. The questionnaires were then mailed, in equal density, to the three geographic regions most affected by the pests (Northeastern United States, Southeastern United States, and the State of Oregon). Each pest management scenario in the questionnaire described the insect, its area of impact, and the effects of an uncontrolled infestation. Three management scenarios were then given and the respondents were asked to rate them on a scale of 1 to 10.

Statistical analysis of the data was conducted using an ordered probit model. The model used the following independent variables: number of forest acres expected to be infested within 15 years of implementation of the management program, cost per household of the management program, expected percent changes in commercial timber harvests, and a dummy variable indicating whether a pest is native or non-native. Haefele and Loomis (2001) found the coefficient on acres infested to be negative and significant, indicating increased disutility for increases in the size of the infested area. The marginal economic value of a one-acre reduction in pest-infested forestland was found to be $0.54 per household.

The authors conclude that households value reduction in forest pest infestations, even in regions far away, which make visitation less likely. Forest infestation reduction is a public good that may affect millions of households. If the passive use value of the infestation reduction is wide spread and in a location with a large population, this value could exceed several millions of dollars for a small reduction.

18. Kramer and others 2003. Contingent Valuation of Forest Ecosystem Protection (Appendix Table 18)

Kramer and others (2003) estimate willingness to pay (WTP) for protection of the remaining healthy spruce-fir forests in the southern Appalachian

Mountains and Great Smokey Mountain National Park. The high elevation spruce-fir forest, covering 26,610 ha of mountain tops in Tennessee, North Carolina, and Virginia, has seen a large increase in spruce-fir mortality since the 1950s. The decline of the spruce-fir forest can easily be seen from roads and trails throughout the area. The cause of this decline is primarily attributed to the balsam woolly adelgid (*Adelges piceae*). The experiment concentrates on two increments of forest protection: areas along roads and trail corridors and the entire ecosystem. Protecting trees along the roads and trails were meant to appeal to people who value the ecosystem for recreational use. Protection of the entire ecosystem was thought to appeal to individuals who value the ecosystem as a whole and not simply the direct benefits it provides.

The authors used a mail-out, mail-back contingent valuation survey to gather WTP estimates for the protection of the remaining spruce-fir forests and other information about the respondents. The survey was sent to households within a 500-mile radius of Ashville, North Carolina, a population with familiarity of the study area. The survey included photographs of three stages of forest decline and a map of the area of interest. All of the questions contained in the survey were in dichotomous choice format. The data collected was used to test three hypotheses: (1) WTP >$0; (2) incremental WTP increases at a decreasing rate; and (3) people would be willing to pay to protect an entire ecosystem. A bivariate probit model was used to estimate the parameters of the explanatory variables and the Krinsky-Robb bootstrap technique was used to test the hypotheses.

The authors, after testing their three hypotheses, found that incremental WTP for ecosystem protection is positive and WTP for forest protection increases at a decreasing rate. The results of the bivariate probit model showed that factors affecting an individual's WTP included household income (a categorical variable), membership to an environmental organization, and if protecting the spruce-fir forest for reasons including recreational activities and scenic beauty were important to them or not. The results for an individual's WTP were broken up into three categories: use value, bequest value, and existence value. Use value, was an estimated $4 (13 percent of total WTP); bequest value was an estimated $8 (30 percent of total WTP) and the estimated existence value was $16 (57 percent of total WTP) Total value for spruce-fir forest protection amounts to $28 per person.

The authors conclude that the assessment of forest values is a useful practice, as the results have many applications in policy and management. Not only can estimates from nonmarket valuation studies be used directly, but they can also improve the understanding of the economic importance of the structure, health, and extent of forest ecosystems. This study, among others, shows that protection and restoration of forest ecosystems is an economic good that people are willing to pay for, and, therefore, it should be valued as such.

19. Asaro and others 2006. Control of Low-Level Nantucket Pine Tip Moth Populations: A Cost Benefit Analysis (Appendix Table 19)

Asaro and others (2006) obtained damage estimates from 200 trees over a 3-year period to estimate the willingness to pay (WTP) for Nantucket pine tip moth (*Rhyacionia frustrana*) control at the beginning of a rotation of loblolly pine. They also aimed to establish an economic injury level that, once reached, justified the application of pesticides. Loblolly pine, *Pinus taeda* L., is the most commercially important tree species in the southeastern United States. The pine tip moth has the largest effect of any insect on the annual growth of loblolly pine seedlings and saplings. The pine tip moth causes significant long-term growth losses in the pines and the effects are especially prevalent in monoculture type environments, such as intensively managed plantations.

The authors used two loblolly pine plantations located in Oglethorpe County, Georgia, to conduct their experiment. The sites were chosen because of the three annual generations of pine tip moth in the region and the availability of non-vegetated land. Each of the two sites was planted with 1,750 trees/ha in 1998. By random design, a plot of 200 trees was established at each site, to maximize degrees of freedom for error. The study was designed to compare trees in different damage categories, thus trees selected for treatment and measurement were positioned 10 per row, every other tree, and every other row. Within the plots, 150 trees were left untreated while 50 trees were sprayed with pesticides three times a year. Damage to each tree was measured at the end of each pine tip moth generation. The relevant measure, percent shoot damage to each tree, was obtained by counting all infested and uninfested shoots on a tree. Pine tree heights and diameters were measured at the end of the 3rd year of the study once growth had ended. The data for each tree was then converted into a volume index.

The authors used the data collected to calculate the total value of pine tip moth control, the difference between the profitability of timber production with and without treatment at each of the separate damage categories (intervals of 10% damage). After estimating bare land values and the cost of planting and fertilization, the authors calculated WTP values at three different interest rates: 3, 5, and 7%. The authors found that there was no difference between sprayed and unsprayed trees with 0-10% damage. "Trees in the 10-20% damage category averaged 28.4 and 16.5% less volume than sprayed trees at sites 1 and 2, respectively" (p. 184). "Trees in the 20-30% damage category averaged 48.2 and 26.2% less volume than sprayed trees for sites 1 and 2, respectively, but only site 1 was statistically significant" (p. 185). In all cases the WTP for control increases as the discount rate declines and damage increases. The estimates for WTP to reduce loblolly pine damage from Nantucket pine tip moth over one rotation are as follows.

Damage category	Discount rate	WTP
0-10%	3%	$586 to $1,482
	5%	$319 to $852
	7%	$183 to $516
10-20%	3%	$921 to 2,530
	5%	$497 to $1,433
	7%	$284 to $857
30-40%	3%	$1,413
	5%	$755
	7%	$429

The WTP estimates have management implications, in the form of damage thresholds. This study suggests "that even at the lowest damage levels (10-20%) and highest real discount rates (7%) there is a WTP value of $183, which is enough to accommodate two to three sprays over a 3-year period" (p. 186).

It is concluded that significant financial losses from pine tip moth attack can be prevented. Due to the prevalence of the pine tip moth in this region of the United States and the small profit margins in forestry, low levels of damage are not alarming. However, tip moth management can be important if damage levels consistently exceed 30% of shoots. Economic benefits from pest control in pulpwood and saw timber plantations can be increased if rotation length of loblolly pine is decreased. Another way to increase growth is to predict when pine tip moth populations will exceed 30% damage and spray once or twice a year for the first 2 to 3 years of growth.

20. Holmes and others 2006. Exotic Forest Insects and Residential Property Values (Appendix Table 20)

Holmes and others (2006) estimate the economic damages to homeowners caused by the hemlock woolly adelgid (*Adelges tsugae*) in Sparta, New Jersey. This study tested the relationship between poor hemlock health and the decrease in residential property values and whether or not the effect extends across parcel boundaries. Eastern hemlock trees are widely distributed throughout the Appalachian Mountains, northeastern United States, the northern Midwest, and parts of Canada. Eastern hemlocks play a central role in the forest ecosystems in which it resides. The eastern hemlock is a long lived species, but once a tree is moderately or severely infested by the woolly adelgid, there is little chance for recovery. Hemlocks produce poor quality wood and are, therefore, rarely used in timber production. The primary values obtained from hemlocks come from their aesthetic value and the ecosystem services they provide.

The authors use data from 3,379 residential property sales in the town of Sparta, New Jersey, between 1992 and 2002. Descriptive statistics of the data include a median house age of 29 years, median lot size of one half acre, and a median sales price of $342,260 (2002 dollars). Land cover and use variables, for three spatial scales (0.1 km, 0.5 km, and 1.0 km) from the center of each parcel, were compiled using satellite data. These characteristics included such data as proximity to shopping services, surrounding forest type, presence of bodies of water, and distance from the closest golf course. The technique of image differencing was used to track the different levels of defoliation caused by the hemlock woolly adelgid. A hedonic property value model was used to assess the effects of hemlock quality on property values. The marginal willingness to pay, by consumers, for hemlock health was derived from the variable hemlock health's contribution to sales prices of the residential properties, using first stage hedonic analysis.

The authors found that parameter estimates on many of the housing characteristics were significant at the 0.01 level or higher. An additional bedroom or bathroom adds 3 – 4% to the value of the house. It was also found that additional acreage increases property values, with lots over 1 acre receiving a price premium. Several parameter estimates for land use variables were also significant, including proximity to a golf course, and lakes or ponds on the lot or adjacent to the property. Parameter estimates on healthy hemlocks were positive and significant in all model specifications, indicating that hemlocks are valued for their unique aesthetic qualities and people enjoy living in or nearby hemlock stands. Thus, hemlock health affects property values where they are located and in the broader neighborhood. Hemlocks with severe decline were not statistically significant. However, dead hemlocks were found to be positive and statistically significant within the 1.0 km buffer from the lot center, suggesting that Sparta residents value the inevitable growth of new trees and vegetation.

Hemlock trees possess aesthetic characteristics that people value. Holmes and others (2006) found that a 1-acre increase in land containing healthy hemlocks increases property value by 0.66 to 8.08%, depending on distance from the property's center. Property value decreases 0.96 to 4.76% when the number of moderately healthy trees increases by 1 acre. The authors found that increases in severely defoliated hemlock trees were statistically insignificant. A 1 acre increase in dead trees, within a 1.0 km buffer of the property center, was found to increase property value by 2.11%; demonstrating that Sparta residents value the regeneration of trees and other vegetation in their far view.

The results of this study show that hemlock decline, due to hemlock woolly adelgid, is connected to the decline of residential property sale prices in Sparta, New Jersey. This phenomenon extends to neighboring properties that also contain declining hemlocks. The authors conclude that the "contribution of healthy hemlock stands to residential property values appears to be qualitatively different than other tree species in the housing market under investigation" (p. 164), which can be attributed to their unique visual beauty. Thus, the landscape externalities caused by the hemlock woolly adelgid should be mitigated in order to preserve home values.

21. Huggett and others 2008. Forest Disturbance Impacts on Residential Property Values (Appendix Table 21)

Huggett and others (2008) examined the economic impacts of the hemlock woolly adelgid on private property values in West Milford, New Jersey. The hedonic property value model was used to test the relationship between hemlock decline and residential property values. Hemlocks are widely used as ornamental trees in residential landscapes, and have shown little or no opposition to attacks by the woolly adelgid. Hemlock trees located near roadways or in small residential

landscapes can be easily treated with insecticides during the early stages of infestation, but once infestation becomes severe, defoliation and loss of tree vigor results in tree death.

The impacts of hemlock woolly adelgid happen gradually over the course of several years. Symptoms begin with moderate decline and eventually end in tree mortality. The authors hypothesized that there exists a threshold, beyond which the presence of hemlocks shifts the property value function. Thus, the authors aimed to find the point in the infestation process at which the hemlock decline registers an impact on property prices. Data were collected on many structural and landscape aspects of the land parcels sold between 1992 and 2002, including total area of hemlock trees. A second hemlock variable was also created to evaluate the point in time at which the hemlocks decline began to affect housing prices. Data on the price of the house and date sold were also collected. The average sale price for the sample was $177,752. Hemlock health data were divided into 4 categories: healthy (<25% defoliation), moderate decline (25-50% defoliation), severe decline (50-75% defoliation), and dead (>75% defoliation).

Results of the OLS regression showed that the data fit the model very well and that several land cover variables play a critical role in determining property values. While mixes of healthy and unhealthy hemlocks were found on the parcels sold during 1992 and 2002, the authors found that hemlock health declined rapidly on parcels sold in 2000 and subsequent years. The parameter estimate for total area of hemlock trees was found to be insignificant for the early stages of the outbreak. This indicates that when hemlock trees are healthy or in moderate decline they do not influence property value. In contrast, the parameter estimate on hemlock forests late in the epidemic was negative and significant. The results show that the marginal effect of 1 additional acre of severely defoliated or dead hemlocks decreased property value by 8.3% during the study period. Because hemlocks are valued for their aesthetic qualities, this loss in value can be attributed to the presence of severely defoliated and dead trees.

22. Price and others 2010. Insect Infestation and Residential Property Values: A Hedonic Analysis of the Mountain Pine Beetle Epidemic (Appendix Table 22)

Price and others (2010) estimate willingness to pay (WTP) to prevent mountain pine beetle, *Dendroctonus ponderosae*, damage on residential property values in

Grand County, Colorado. Over the last 15 years (1996 to 2010) mountain pine beetle, along with several other species of bark beetles, have been a destructive force in the western United States and Canada. Recent studies have shown that the large pine beetle outbreaks can be attributed to the drought conditions brought on by a changing climate and forest management practices that have suppressed fire over the last century. The most prominent infestations have taken place in Colorado and British Columbia. Agencies such as the Colorado State Forest Service and the British Columbia Ministry of Forests have estimated the scale of the mountain pine beetle infestation, amounting to approximately 14 million hectares between 1996 and 2007. However, an important outcome of the mountain pine beetle damage is a loss in utility, derived from goods and services, for residents of the wildland-urban interface (WUI) through reduced forest amenities and increased wildfire risk.

There are many benefits to living on the WUI including recreation opportunities, scenic views, and increased property values. The mountain pine beetle, while integral to the forest ecosystems of Colorado and British Columbia, is capable of reaching epidemic levels and causing high levels of tree mortality. Once an outbreak begins there is very little that can be done to prevent its spread. Trees that are killed by the beetles increase the risk of wildfire and become hazard trees for recreators. In cases of outbreak levels, as in Colorado and British Columbia, the most effective method to control infestation is to eliminate the underlying causes. Usually this means relaxing fire suppression practices, and in areas around the WUI using silviculture techniques such as thinning and controlled burning. The objective of these practices is to return the forests to a more natural state, improve tree health, diversify tree growth stages, and reduce density. Forests that exhibit these qualities are much less vulnerable to mountain pine beetle attacks and wildfire than those that do not. Residents along the WUI in Colorado have attempted to control the outbreak of beetles near their homes by using chemical (e.g., pesticides) and physical (e.g., burning and stripping bark) means. These methods are effective, but only when the populations of beetles are small and well below outbreak levels.

The authors use the hedonic pricing model in Grand County, Colorado, where a vast majority of the county's population lives in the WUI and there exists one of the highest concentrations of lodgepole pine forests in Colorado. As a result, a large portion of the county's forests have been profoundly affected by the mountain pine beetle. Grand County is home to two large reservoirs, multiple wilderness areas, a portion of Rocky Mountain National Park, and two ski resorts; hence, their economy is highly dependent on the recreation and nature-based tourism industries. The data for analysis included the sales prices and housing characteristics of homes sold between 1995 and 2006, land use maps from the U.S. Geological Survey, and data on mountain pine beetle infestations (annual aerial detection surveys) provided by the U.S. Forest Service.

The results showed that housing prices are negatively correlated with the number of trees killed by mountain pine beetle. All estimated coefficients were significant at the 0.01 level for each of the three buffer zones (0.1 km, 0.5 km and 1.0 km radius from the home) and increased as the radius of the buffer zone increased; indicating the closer a damaged tree is to the house the greater the impact it will have on utility and price. Homeowners' marginal WTP to avoid utility loss from mountain pine beetle damage within 0.1 km radius of a home was estimated to be approximately $648 per tree. Similarly, within 0.5 km and 1.0 km radiuses the marginal WTP estimates were approximately $43 and $17 per tree respectively.

The authors conclude that the combination of drought, overcrowded forests, and the mountain pine beetle epidemic indicate that the way forests are currently managed in Colorado is damaging the ecosystem. They also argue that the continued human actions that invite mountain pine beetle infestations reduce the stock of natural capital for future generations, at least in the short term. Forest restoration provides a range of social benefits obtained by homeowners and recreationists, both on and off site, as well as non-use values. This study suggests that an appropriate solution to restore forest health, minimize wildfire risk, and address the beetle problem may be to establish public programs to maintain healthy forests. It also suggests that partial funding for forest preservation is available from WUI residents through the form of taxation or cost share programs.

Appendix C: Summary Tables of Nonmarket Economic Studies of Forest Insect Pest Damages

Appendix C Table 1—Payne and others 1973. Economic analysis of the gypsy moth problem in theNortheast: II. Applied to residential property.

Category	Description
Region, forest type	Pennsylvania/Maryland, mixed hardwood on residential properties
Stakeholder	Homeowners
Insect	Gypsy moth (*Lymantria dispar*)
Origin	Non-native
Indicator variable effect	Decrease in no. of trees 6 in. dbh
Value type	Contributed residential property value of trees
Valuation method, estimate type	Hedonic pricing, hedonic price/acre or per tree
Estimated value	$1,175/acre or $270/tree loss with 15-pct decrease in number of trees

Appendix C Table 2—Wickman and Renton 1975. Evaluating damage caused to a campground by Douglas-fir tussock moth.

Category	Description
Region, forest type	Stowe Reservoir campground in California, white fir
Stakeholder	Recreationists
Insect	Douglas-fir tussock moth (*Orgyia pseudotsugata*)
Origin	Native
Indicator variable effect	Decrease in no. of trees in campground
Value type	Recreation, aesthetic
Valuation method, estimate type	Replacement and clean-up costs allocation, cost/tree
Estimated value	$56 aesthetic value/tree, $13 clean-up cost/dead tree, with 25trees killed total damage to campground is $1,725

Appendix C Table 3—Michalson 1975. Economic impact of mountain pine beetle on outdoor recreation.

Category	Description
Region, forest type	Island Park area in Targhee National Forest (Idaho), lodgepole pine
Stakeholder	Recreationists
Insect	Mountain pine beetle (*Dendroctonus ponderosae*)
Origin	Native
Indicator variable effect	Increase in no. of visible dead trees
Value type	Recreation
Valuation method, estimate type	Travel cost, consumer surplus/person-expenditure/person visitor days/person
Estimated value	$4.09 consumer surplus, $0.43 expenditure, and 1.2 days with a>30-pct infestation

Appendix C Table 4—Moeller and others, 1977. Economic analysis of the gypsy moth problem in the Northeast: III. Impacts on homeowners and managers of recreation areas.

Category	Description
Region, forest type	New York/Pennsylvania, northeastern deciduous
Stakeholder	(a)[1] homeowners
	(b)[1] managers of recreation areas
Insect	Gypsy moth (*Lymantria dispar*)
Origin	Non-native
Indicator variable effect	Presence of moth, defoliation, tree mortality
Value type	(i)[2] control cost
	(ii)[2] financial loss
	(iii)[2] recreation loss
Valuation method, estimate type	(i, ii, iii) sample average, (a-i, ii) annual public and commercial control costs and financial loss/household, (b-i, ii) annual control costs or financial loss for commercial and quasi-public campgrounds, (a-iii) annual person-days per household, (b-iii) annual person-days per campground
Estimated value	(a-i) $102 for public control, $240 for commercial control cost;
	(a-ii) $125 with public control, $479 for commercial control financial loss;
	(a-iii) 108 person-days with public control, 133 person-days with commercial control recreation loss;
	(b-i) $441 for commercial campgrounds, $722 for quasi-public campgrounds control costs;
	(b-ii) $249 for commercial campgrounds, $996 for quasi-public campgrounds in financial loss;
	(b-iii) 161 person-days for commercial, 240 person-days for quasi-public, 36,660 person-days for public campgrounds per unit in recreation loss

[1](a) and (b) refer to corresponding stakeholder.

[2](i), (ii), and (iii) refers to corresponding value type.

Appendix C Table 5—Leuschner and Young 1978. Estimating the southern pine beetle's impact on reservoir campsites

Category	Description
Region, forest type	East Texas Forest Service and Corps of Engineers campgrounds, mixed pine and hardwood
Stakeholder	Recreationists
Insect	Southern pine beetle (*Dendroctonus frontalis*)
Origin	Native
Indicator variable effect	Decrease in percent pine crown cover
Value type	Contributed recreation value of pine crown cover
Valuation method, estimate type	Travel cost, consumer surplus/person/visit
Estimated value	$3.37 for Forest Service campgrounds, $2.44 for Corps of Engineers campgrounds per person with a 30-pct reduction in pine crown cover

Appendix C Table 6—Walsh and Olienyk 1981. Recreation demand effects of mountain pine beetle damage to the quality of forest recreation resources in the Colorado Front Range.

Category	Description
Region, forest type	Colorado Front Range, 6,000-8,000 ft elevation in Rocky Mountains, mixed-age ponderosa pine
Stakeholder	Recreationists
Insect	Mountain pine beetle (*Dendroctonus ponderosae*)
Origin	Native
Indicator variable effect	(a)[1] decrease in no. of trees ≥6 in. dbh/acre
	(b)[1] decrease in average tree size
	(c)[1] increase in pct visible tree damage
	(d)[1] increase in pct dead and down trees on ground with slash
	(e)[1] increase in pct pest-caused treeless areas in acres
	(f)[1] decrease in no. of large tree ≥24 in. dbh
Value type	Contributed recreation value of indicator variable
Valuation method, estimate type	(i)[2] contingent valuation, annual willingness to participate in user-days
	(ii)[2] contingent valuation, consumer surplus/user-day
	(iii)[2] travel cost, no. of trips/person
	(iv)[2] travel cost, annual consumer surplus/person and per trip
Estimated value	(i) 370,000 user-days loss with 15-pct decrease in (a) on-site, 330,000 user-days loss with 15-pct decrease in (a) in nearview, 207,000 user-days loss with 15-pct decrease in (a) in farview, 422,000 user-days loss with 15-pct decrease in (b), 3,045,000 user-days loss with 15-pct increase in (c) or (d), 317,000 user-days loss with 15-pct increase in (e), 192,700 user-days loss with 15-pct decrease in (f);
	(ii) $1.50/user-day loss at 178 trees/acre, or $1.70/user-day loss at 270 trees/acre with 15-pct decrease in (a);
	(iii) 0.16 fewer trips per person with 15-pct decrease in (a);
	(iv) $11.60/person, or $1.75/person/trip with 15-pct decrease in (a)

[1](a), (b), (c), (d), (e), and (f) correspond to indicator variable effect.

[2](i), (ii), (iii), and (iv) correspond to valuation method and estimate type.

Appendix C Table 7—Walsh and others 1981a. Value of trees to residential property owners with mountain pine beetle and spruce budworm damage in the Colorado Front Range.

Category	Description
Region, forest type	Colorado Front Range, 6,000-8,000 ft elevation in Rocky Mountains, mixed-age ponderosa pine
Stakeholder	Homeowners, property owners
Insect	Mountain pine beetle (*Dendroctonus ponderosae*), western spruce budworm (*Choristoneura occidentalis*)
Origin	Both native
Indicator variable effect	(a)[1] decrease in no. of trees ³6 in. dbh/acre
	(b)[1] decrease in average tree size
	(c)[1] expectation of 50-pct tree loss
	(d)[1] increase in pct visible tree damage
	(e)[1] increase in pct dead and down trees on ground with slash
	(f)[1] increase in pct pest-caused treeless areas in acres
	(g)[1] decrease in no. of large tree ³24 in. dbh
Value type	Contributed property value of indicator variable
Valuation method, estimate type	Contingent valuation, annual willingness to pay
Estimated value	$984/acre impr. lots, $578/acre unimpr.[2] lots loss with 15-pct decrease in (a) on-site;
	$602/acre impr. lots loss with 15-pct decrease in (a) on adjacent lots in near view;
	$1,228/acre impr. lots, $783/acre unimpr. lots loss with 15-pct decrease in (b);
	$2,351/acre impr. lots, $1,364/acre unimpr. lots loss with (c);
	$7,034/acre impr. lots, $4,045/acre unimpr. lots loss with 15-pct increase in (d);
	$907/acre impr. lots, $896/acre unimpr. lots loss with 15-pct increase in (e);
	$61/acre impr. lots, $84/acre unimpr. lots gain with 15-pct increase in (f);
	$918/acre impr. lots, $449/acre unimpr. lots loss with 15-pct decrease in (g)

[1](a), (b), (c), (d), (e), (f), and (g) correspond to indicator variable effect.

[2]Improved lots (lots with buildings, primarily residences) is abbreviated "impr.," while unimproved lots (lots with no buildings) is abbreviated "unimpr."

Appendix C Table 8—Walsh and others 1981b. Appraised market value of trees on residential property with mountain pine beetle and spruce budworm damage in the Colorado Front Range.

Category	Description
Region, forest type	Colorado Front Range, 6,000-8,000 ft elevation in Rocky Mountains, mixed-age ponderosa pine
Stakeholder	Real estate appraisers, homeowners, property owners
Insect	Mountain pine beetle (*Dendroctonus ponderosae*), western spruce budworm (*Choristoneura occidentalis*)
Origin	Both native
Indicator variable effect	(a)[1] decrease in no. of trees $\geq$6 in. dbh/acre
	(b)[1] decrease in average tree size
	(c)[1] expectation of 50-pct tree loss
	(d)[1] increase in pct visible tree damage
	(e)[1] increase in pct dead and down trees on ground with slash
	(f)[1] increase in pct pest-caused treeless areas in acres
	(g) decrease in no. of large tree $\geq$24 in. dbh
Value type	Contributed property value of indicator variable
Valuation method, estimate type	Contingent valuation, appraised market value
Estimated value	$209/acre impr. lots, $201/acre unimpr.[2] lots loss with 15-pct decrease in (a) on-site;
	$241/acre impr. lots loss with 15-pct decrease in (a) on adjacent lots in near view;
	$1,155/acre impr. lots, $1,014/acre unimpr. lots loss with 15-pct decrease in (b);
	$1,810/acre impr. lots, $1,492/acre unimpr. lots loss with (c);
	$3,902/acre impr. lots, $2,715/acre unimpr. lots loss with 15-pct increase in (d) on-site;
	$1,688/acre impr. lots, $2,445/acre unimpr. lots loss with 15-pct increase in (d) on adjacent lots in near-view;
	$259/acre impr. lots, $848/acre unimpr. lots loss with 15-pct increase in (d) in far-view;
	$493/acre impr. lots, $476/acre unimpr. lots loss with 15-pct increase in (e);
	$102/acre impr. lots, $102/acre unimpr. lots gain with 15-pct increase in (f);
	$251/acre impr. lots, $294/acre unimpr. lots loss with 15-pct decrease in (g)

[1](a), (b), (c), (d), (e), (f), and (g) correspond to indicator variable effect.

[2]Improved lots (lots with buildings, primarily residences) is abbreviated "impr.," while unimproved lots (lots with no buildings) is abbreviated "unimpr."

Appendix C Table 9—Loomis and Walsh 1988. Recreation and tree stand characteristics in the Colorado Front Range.

Category	Description
Region, forest type	Colorado Front Range, 6,000-8,000 ft elevation in Rocky Mountains, mixed-age ponderosa pine
Stakeholder	Recreationists
Insect	Mountain pine beetle *(Dendroctonus ponderosae)*, western spruce budworm *(Choristoneura occidentalis)*
Origin	Both native
Indicator variable effect	(a)[1] decrease in no. of trees ≥6 in. dbh
	(b)[1] decrease in average tree size
Value type	Contributed recreation value of indicator variable
Valuation method, estimate type	Contingent valuation, annual willingness to pay
Estimated value	(a) $145 benefit/visitor for camping with 200 trees/acre; $169 benefit/visitor for picnicking with 200 trees/acre; $161 benefit/visitor for backpacking with 200 trees/acre; $302 benefit/visitor for hiking with 200 trees/acre; $321 benefit/ visitor for fishing with 200 trees/acre; $97 benefit/visitor for off-roading with 200 trees/acre;
	(b) $210 annual benefit/visitor with avg. tree size at 10.5 in. dbh, $28 annual benefit/visitor with avg. tree size at 2.5 in. dbh; $13 annual benefit/visitor/day with avg. tree size at 13 in. dbh; $5 annual benefit/visitor/day with avg. tree size at 4 in. dbh

[1](a) and (b) correspond to indicator variable.

Appendix C Table 10—Walsh and others 1989. Net economic benefits of recreation as a function of tree stand density.

Category	Description
Region, forest type	Colorado Front Range, 6,000-8,000 ft elevation in Rocky Mountains, mixed-age ponderosa pine
Stakeholder	Recreationists
Insect	Mountain pine beetle *(Dendroctonus ponderosae)*
Origin	Native
Indicator variable effect	Decrease in no. of trees ≥6 in. dbh/acre
Value type	Contributed recreation value of indicator variable
Valuation method, estimate type	(i)[1] contingent valuation, annual willingness to pay/person/day
	(ii)[1] travel cost, annual consumer surplus/person/day
Estimated value	(i) $24 net average benefit with 178 trees/acre and 2.7 days/trip;
	(ii) $20 to $26 net average benefit with 178 trees/acre and 2.7 days/trip

[1](i) and (ii) correspond to valuation method and estimate type.

Appendix C Table 11—Walsh and others 1990. Estimating the public benefits of protecting forest quality.

Category	Description
Region, forest type	Colorado National Forests, mixed tree stands
Stakeholder	Recreationists
Insect	Mountain pine beetle *(Dendroctonus ponderosae)*, western spruce budworm *(Choristoneura occidentalis)*
Origin	Both native
Indicator variable effect	Decrease in no. of trees $\geq$6 in. dbh/acre
Value type	Recreation, bequest, existence, option
Valuation method, estimate type	Contingent valuation, annual willingness to pay/household
Estimated value	$52 average annual total value for 150 trees/acre where $14 is for recreation-use, $16 is for bequest, $11 is for existence, and $11 is for option

Appendix C Table 12—Jakus and Smith 1991. Measuring use and nonuse values for landscape amenities: a contingent behavior analysis of gypsy moth control.

Category	Description
Region, forest type	Southcentral Pennsylvania and northcentral Maryland, urban residential, parks, and greenways
Stakeholder	Homeowners
Insect	Gypsy moth *(Lymantria dispar)*
Origin	Non-native
Indicator variable effect	Decrease in aesthetic quality (pct defoliation)
Value type	Aesthetics
Valuation method, estimate type	Contingent valuation, annual willingness to pay/household
Estimated value	$254 to $534 for a private control program, $314 to $670 for a public control program

Appendix C Table 13—Haefele and others 1992. Estimating the total value of forest quality in high elevation spruce-fir forests.

Category	Description
Region, forest type	4,400 to 6,684 ft elevation in Appalachian Mountains of North Carolina, Tennessee, and Virginia, spruce-fir forests
Stakeholder	Recreationists, general public
Insect	Balsam woolly adelgid *(Adelges piceae)*
Origin	Non-native
Indicator variable effect	Increase in perceived visible damage (dead and dying trees)
Value type	Recreation, bequest, existence, and option
Valuation method, estimate type	Contingent valuation, annual willingness to pay/household
Estimated value	$19 to $63 for forests near roads and trails, $22 to $107 for total forest

Appendix C Table 14—Miller and Lindsay 1993. Willingness to pay for a state gypsy moth control program in New Hampshire: a contingent valuation case study.

Category	Description
Region, forest type	New Hampshire, northeastern deciduous forest
Stakeholder	New Hampshire residents
Insect	Gypsy moth (*Lymantria dispar*)
Origin	Non-native
Indicator variable effect	Increase in perceived visible damage
Value type	Total value
Valuation method, estimate type	Contingent valuation, annual willingness to pay/household
Estimated value	$70 average annual, $43 median, or mean $16/acre, median $10/acre

Appendix C Table 15—Holmes and Kramer 1996. Contingent valuation of ecosystem health.

Category	Description
Region, forest type	4,400 to 6,684 ft elevation in Appalachian Mountains of North Carolina, Tennessee, and Virginia, spruce-fir forests
Stakeholder	Recreationists, general public
Insect	Balsam woolly adelgid (*Adelges piceae*)
Origin	Non-native
Indicator variable effect	Increase in perceived visible damage (dead and dying trees)
Value type	Existence and total
Valuation method, estimate type	Contingent valuation, annual willingness to pay/household
Estimated value	$11 median existence value, $36 median total value

Appendix C Table 16—Thompson and others 1999. Valuation of tree aesthetics on small urban-interface properties.

Category	Description
Region, forest type	Lake Tahoe Basin, California/ Nevada, Sierra-Nevada mixed conifer and Jeffrey pine forests
Stakeholder	Homeowners
Insect	Unidentified
Origin	Unknown
Indicator variable effect	Area weighted average of the average infection rating by plant group
Value type	Influence of tree care on property value
Valuation method, estimate type	Hedonic pricing, hedonic price/ unit of infected trees
Estimated value	$26,390/unit, if infected trees are thinned then the value of the house will increase by 5 to 30% in value

Appendix C Table 17—Haefele and Loomis 2001. Using the conjoint analysis technique for the estimation of passive use values of forest health.

Category	Description
Region, forest type	Northeast, ornamental and trees in recreational areas; Northwest, fir stands and commercial timber; Southeast, commercial timber and wilderness areas
Stakeholder	General public
Insect	Gypsy moth (*Lymantria dispar*), western spruce budworm (*Choristoneura occidentalis*), southern pine beetle (*Dendroctonus frontalis*)
Origin	Non-native, native, and native, respectively
Indicator variable effect	Forest acreage infested after 15 years
Value type	Passive-use value
Valuation method, estimate type	Contingent choice, price/acre to reduce infestation
Estimated value	$0.54/ acre/ person to reduce infestation

Appendix C Table 18—Kramer and others 2003. Contingent valuation of forest ecosystem protection.

Category	Description
Region, forest type	Southern Appalachian Mountains, Virginia/ North Carolina/ Tennessee, high elevation spruce-fir forest ecosystems
Stakeholder	General public
Insect	Balsam woolly adelgid (*Adelges piceae*)
Origin	Non-native
Indicator variable effect	Presence of the insect along road and trail corridors and presence of the insect in the entire forest ecosystem
Value type	Total value
Valuation method, estimate type	Contingent valuation, willingness to pay for forest ecosystem protection
Estimated value	Use value $4/person; Existence value $16/person; Bequest value $8/person; Total value $28/person

Appendix C Table 19—Asaro and others 2006: Control of low-level Nantucket pine tip moth populations: A cost benefit analysis.

Category	Description						
Region, forest type	Oglethorpe County, Georgia, Loblolly pine plantations						
Stakeholder	Land managers; plantation managers						
Insect	Nantucket pine tip moth (*Rhyacionia frustrana*)						
Origin	Native						
Indicator variable effect	Visible shoot damage after each moth generation and tree volume after 3-year study, in sprayed and unsprayed tree groups						
Value type	Total value of pine tip moth control						
Valuation method, estimate type	Cost benefit analysis, total value of control/percent damage for a given interest rate						
		Site 1			Site 2		
Estimated value	Interest Rate	3%	5%	7%	3%	5%	7%
	10–20% Damage	$1,482	$852	$516	$586	$319	$183
	20–30% Damage	$2,530	$1,433	$857	$921	$497	$284
	30–40% Damage	NA	NA	NA	$1,413	$755	$429

Appendix C Table 20—Holmes and others 2006. Exotic forest insects and residential property values.

Category	Description
Region, forest type	Sparta, New Jersey, individual hemlock trees and hemlock stands on residential properties
Stakeholder	Homeowners
Insect	Hemlock woolly adelgid (*Adelges tsuga*)
Origin	Non-native
Indicator variable effect	Four levels of hemlock defoliation: healthy < 25% defoliation, moderate 25-50% defoliation, severe 50-75% defoliation, and dead
Value type	Contributed residential property value of trees
Valuation method, estimate type	Hedonic property, hedonic price/acre for a level of hemlock defoliation
Estimated value	Healthy: 0.66-8.08% increase in property value/ additional acre of hemlocks
	Moderate: 0.96-4.76% increase in property value/ additional acre of hemlocks
	Severe: inadequate data
	Dead: 2.11% increase in property value/ additional acre of hemlocks

Appendix C Table 21—Huggett and others 2008. Forest disturbance impacts on residential property values.

Category	Description
Region, forest type	West Milford, New Jersey, individual hemlock trees and hemlock stands on residential properties
Stakeholder	Homeowners
Insect	Hemlock woolly adelgid (*Adelges tsuga*)
Origin	Non-native
Indicator variable effect	Four levels of hemlock defoliation: healthy < 25% defoliation, moderate 25-50% defoliation, severe 50-75% defoliation, and dead
Value type	Contributed residential property value of trees
Valuation method, estimate type	Hedonic property, hedonic price/acre for a level of hemlock defoliation
Estimated value	Healthy: 0.66-8.08% increase in property value/ additional acre of hemlocks
	Moderate: 0.96-4.76% increase in property value/ additional acre of hemlocks
	Severe: inadequate data
	Dead: 2.11% increase in property value/ additional acre of hemlocks

Appendix C Table 22—Price and others 2010. Insect infestation and residential property values: A hedonic analysis of the mountain pine beetle epidemic.

Category	Description
Region, forest type	Grand County, Colorado, coniferous forests that provide goods and services to residents of the wildland-urban interface
Stakeholder	Homeowners
Insect	Mountain pine beetle (*Dendroctonus ponderosae*)
Origin	Native
Indicator variable effect	Number of trees killed by the mountain pine beetle
Value type	Contributed property value of forested areas
Valuation method, estimate type	Hedonic property, marginal implicit price/killed tree within a Xkm radius of a home
Estimated value	$648/killed tree within 0.1km of a home
	$43/ killed tree within 0.5km of a home
	$17/ killed tree within 1.0km of a home

The Rocky Mountain Research Station develops scientific information and technology to improve management, protection, and use of the forests and rangelands. Research is designed to meet the needs of the National Forest managers, Federal and State agencies, public and private organizations, academic institutions, industry, and individuals. Studies accelerate solutions to problems involving ecosystems, range, forests, water, recreation, fire, resource inventory, land reclamation, community sustainability, forest engineering technology, multiple use economics, wildlife and fish habitat, and forest insects and diseases. Studies are conducted cooperatively, and applications may be found worldwide.

Station Headquarters
Rocky Mountain Research Station
240 W Prospect Road
Fort Collins, CO 80526
(970) 498-1100

Research Locations

Flagstaff, Arizona	Reno, Nevada
Fort Collins, Colorado	Albuquerque, New Mexico
Boise, Idaho	Rapid City, South Dakota
Moscow, Idaho	Logan, Utah
Bozeman, Montana	Ogden, Utah
Missoula, Montana	Provo, Utah

www.fs.fed.us/rmrs